SCHAUM'S *Easy* OUTLINES

LINEAR ALGEBRA

BASED ON SCHAUM'S
Outline of Theory and Problems of Linear Algebra, Third Edition

BY SEYMOUR LIPSCHUTZ, Ph.D.
AND MARC LARS LIPSON, Ph.D.

ABRIDGEMENT EDITOR:
KIMBERLY S. KIRKPATRICK, Ph.D.

SCHAUM'S OUTLINE SERIES

McGRAW-HILL

New York Chicago San Francisco Lisbon London Madrid Mexico City Milan New Delhi San Juan Seoul Singapore Sydney Toronto

SEYMOUR LIPSCHUTZ is on the faculty of Temple University and formerly taught at the Polytechnic Institute of Brooklyn. He received his Ph.D. in 1960 at Courant Institute of Mathematical Sciences of New York University. He is one of Schaum's most prolific authors and has written, among others, Schaum's Outlines of Beginning Linear Algebra, Probability, Discrete Mathematics, Set Theory, Finite Mathematics, and General Topology.

MARC LARS LIPSON teaches at the University of Georgia. He received his Ph.D. in finance in 1994 from the University of Michigan, and he is the co-author with Seymour Lipschutz of Schaum's Outlines of Discrete Mathematics and Probability.

KIMBERLY S. KIRKPATRICK teaches mathematics at Transylvania University in Lexington, Kentucky. She earned a B.S. in Mathematics Education, a Master of Applied Mathematics, and a Ph.D. in Mathematics, all from Auburn University. She is the co-author of professional papers and previously taught at the University of Evansville. She was the abridgement editor of *Schaum's Easy Outline: Precalculus*.

1 2 3 4 5 6 7 8 9 DOC DOC 0 9 8 7 6 5 4 3 2

ISBN 0-07-139880-5

Library of Congress Cataloging-in-Publication Data applied for.

Sponsoring Editor: Barbara Gilson
Production Supervisors: Tama Harris and Clara Stanley
Editing Supervisor: Maureen B. Walker

*A Division of The **McGraw-Hill** Companies*

Contents

Chapter 1
VECTORS IN $\mathbf{R}^n$

IN THIS CHAPTER:

- ✔ *Vectors in* $\mathbf{R}^n$
- ✔ *Vector Addition and Scalar Multiplication*
- ✔ *Dot Product*

Vectors in $\mathbf{R}^n$

Although we will restrict ourselves in this chapter to vectors whose elements come from the field of real numbers, denoted by **R**, many of our operations also apply to vectors whose entries come from some arbitrary field K. In the context of vectors, the elements of our number fields are called *scalars*.

Lists of Numbers

Suppose the weights (in pounds) of eight students are listed as follows:

$$156 \quad 125 \quad 145 \quad 134 \quad 178 \quad 145 \quad 162 \quad 193$$

One can denote all the values in the list using only one symbol, say w, but with different subscripts; that is

$$w_1 \quad w_2 \quad w_3 \quad w_4 \quad w_5 \quad w_6 \quad w_7 \quad w_8$$

Observe that each subscript denotes the position of the value in the list. For example,

$$w_1 = 156, \text{ the first number, } w_2 = 125, \text{ the second number,} \ldots$$

Such a list of values, $w = (w_1, w_2, w_3, \ldots, w_8)$ is called a *linear array* or *vector*.

The set of all n-tuples of real numbers, denoted by $\mathbf{R}^n$, is called *n-space*. A particular n-tuple in $\mathbf{R}^n$, say $u = (a_1, a_2, \ldots, a_n)$ is called a *point* or *vector*. The numbers a_i are called the *coordinates*, *components*, *entries*, or *elements* of u. Moreover, when discussing the space $\mathbf{R}^n$, we use the term *scalar* for the elements of $\mathbf{R}$.

Two vectors, u and v, are *equal*, written $u = v$, if they have the same number of components and if the corresponding components are equal. Although the vectors (1, 2, 3) and (2, 3, 1) contain the same three numbers, these vectors are not equal since corresponding entries are not equal.

The vector $(0, 0, \ldots, 0)$ whose entries are all 0 is called the *zero vector*, and is usually denoted by 0.

Example 1.1.

(*a*) The following are vectors:

$$(2, -5) \quad (7, 9) \quad (0, 0, 0) \quad (3, 4, 5)$$

The first two belong to $\mathbf{R}^2$ whereas the last two belong to $\mathbf{R}^3$. The third is the zero vector in $\mathbf{R}^3$.

(*b*) Find x, y, z such that $(x - y, x + y, z - 1) = (4, 2, 3)$.

By definition of equality of vectors, corresponding entries must be equal. Thus,

$$x - y = 4 \quad x + y = 2 \quad z - 1 = 3$$

Solving this system of equations yields $x = 3$, $y = -1$, $z = 4$.

Column Vectors

Sometimes a vector in n-space $\mathbf{R}^n$ is written vertically, rather than horizontally. Such a vector is called a *column vector*, and, in this context, the above horizontally written vectors are called *row vectors*. For example, the following are column vectors with 2, 2, 3, and 3 components, respectively:

$$\begin{bmatrix} 1 \\ 2 \end{bmatrix}, \quad \begin{bmatrix} 3 \\ -4 \end{bmatrix} \quad \begin{bmatrix} 1 \\ 5 \\ -6 \end{bmatrix} \quad \begin{bmatrix} 1.5 \\ \frac{2}{3} \\ -15 \end{bmatrix}$$

We also note that any operation defined for row vectors is defined analogously for column vectors.

Vector Addition and Scalar Multiplication

Consider two vectors u and v in $\mathbf{R}^n$, say

$$u = (a_1, a_2, \ldots, a_n) \text{ and } v = (b_1, b_2, \ldots, b_n)$$

Their *sum*, written $u + v$, is the vector obtained by adding corresponding components from u and v. That is,

$$u + v = (a_1 + b_1, a_2 + b_2, \ldots, a_n + b_n)$$

The *scalar product* or, simply, *product*, of the vector u by a real number k, written ku, is the vector obtained by multiplying each component of u by k. That is,

$$ku = k(a_1, a_2, \ldots, a_n) = (ka_1, ka_2, \ldots, ka_n)$$

Observe that $u + v$ and ku are also vectors in $\mathbf{R}^n$. The sum of vectors with different numbers of components is not defined.

Negatives and subtraction are defined in $\mathbf{R}^n$ as follows:

$$-u = (-1)\,u \quad \text{and} \quad u - v = u + (-v)$$

The vector $-u$ is called the negative of u, and $u - v$ is called the *difference* of u and v.

Now suppose we are given vectors $u_1, u_2, \ldots, u_m$ in $\mathbf{R}^n$ and scalars $k_1, k_2, \ldots, k_m$ in $\mathbf{R}$. We can multiply the vectors by the corresponding scalars and then add the resultant scalar products to form the vector

$$v = k_1u_1 + k_2u_2 + k_3u_3 + \ldots + k_mu_m$$

Such a vector v is called a *linear combination* of the vectors $u_1, u_2, \ldots, u_m$.

Example 1.2.
(*a*) Let $u = (2, 4, -5)$ and $v = (1, -6, 9)$. Then

$$\begin{aligned} u + v &= (2 + 1, 4 + (-5), -5 + 9) = (3, -1, 4) \\ 7u &= (7(2), 7(4), 7(-5)) = (14, 28, -35) \\ -v &= (-1)(1, -6, 9) = (-1, 6, -9) \\ 3u - 5v &= (6, 12, -15) + (-5, 30, -45) = (1, 42, -60) \end{aligned}$$

(*b*) Let $u = \begin{bmatrix} 2 \\ 3 \\ -4 \end{bmatrix}, v = \begin{bmatrix} 3 \\ -1 \\ -2 \end{bmatrix}$. Then $2u - 3v = \begin{bmatrix} 4 \\ 6 \\ -8 \end{bmatrix} + \begin{bmatrix} -9 \\ 3 \\ 6 \end{bmatrix} = \begin{bmatrix} -5 \\ 9 \\ -2 \end{bmatrix}$

Basic properties of vectors under the operations of vector addition and scalar multiplication are described in the following theorem.

Theorem 1.1: For any vectors u, v, w in $\mathbf{R}^n$ and any scalars k, k' in $\mathbf{R}$,

(i) $(u + v) + w = u + (v + w)$,
(ii) $u + 0 = u$,
(iii) $u + (-u) = 0$,
(iv) $u + v = v + u$,
(v) $k(u + v) = ku + kv$,
(vi) $(k + k')u = ku + k'u$,
(vii) $(k\,k')u = k(k'u)$,
(viii) $1u = u$.

Suppose u and v are vectors in $\mathbf{R}^n$ for which $u = kv$ for some nonzero scalar k in $\mathbf{R}$. Then u is called a *multiple* of v. Also, u is said to be the *same* or *opposite direction* as v accordingly as $k > 0$ or $k < 0$.

Dot Product

Consider arbitrary vectors u and v in $\mathbf{R}^n$; say,

$$u = (a_1, a_2, \ldots, a_n) \text{ and } v = (b_1, b_2, \ldots, b_n)$$

The *dot product* or *inner product* or *scalar product* of u and v is denoted and defined by $u \cdot v = a_1b_1 + a_2b_2 + \ldots + a_nb_n$. That is $u \cdot v$ is obtained by multiplying corresponding components and adding the resulting products. The vectors u and v are said to be *orthogonal* (or *perpendicular*) if their dot product is zero, that is, if $u \cdot v = 0$.

Example 1.3. Let $u = (1, -2, 3)$, $v = (4, 5, -1)$, $w = (2, 7, 4)$. Then:

$$u \cdot v = 1(4) - 2(5) + 3(-1) = 4 - 10 - 3 = -9$$

and

$$u \cdot w = 1(2) - 2(7) + 3(4) = 2 - 14 + 12 = 0$$

Thus u and w are orthogonal.

Basic properties of the dot product in $\mathbf{R}^n$ follow:

Theorem 1.2: For any vectors u, v, w in $\mathbf{R}^n$ and any scalar k in $\mathbf{R}$:

(i) $(u + v) \cdot w = u \cdot w + v \cdot w$
(ii) $(ku) \cdot v = k(u \cdot v)$
(iii) $u \cdot v = v \cdot u$
(iv) $u \cdot u \geq 0$ and $u \cdot u = 0$ if $u = 0$.

Note that (ii) says that we can "take k out" from the first position in an inner product. By (iii) and (ii), $u \cdot (kv) = (kv) \cdot u = k(v \cdot u) = k(u \cdot v)$ That is, we can also "take k out" from the second position in an inner product.

The space $\mathbf{R}^n$ with the above operations of vector addition, scalar multiplication, and dot product is usually called *Euclidean n-space*.

Norm (Length) of a Vector

The *norm* or *length* of a vector u in $\mathbf{R}^n$, denoted by $\|u\|$, is defined to be the nonnegative square root of $u \cdot u$. In particular, if $u = (a_1, a_2, \ldots, a_n)$, then

$\|u\| = \sqrt{u \cdot u} = \sqrt{a_1^2 + a_2^2 + \cdots + a_n^2}$. That is, $\|u\|$ is the square root of the sum of the squares of the components of u. Thus $\|u\| \geq 0$, and $\|u\| = 0$ if and only if $u = 0$.

A vector u is called a *unit vector* if $\|u\| = 1$ or, equivalently, if $u \cdot u = 1$.

For any nonzero vector v in $\mathbf{R}^n$, the vector $\hat{v} = \dfrac{1}{\|v\|} v = \dfrac{v}{\|v\|}$ is the unique unit vector in the same direction as v. The process of finding $\hat{v}$ from v is called *normalizing* v.

Example 1.4. Suppose $u = (1, -2, -4, 5, 3)$. To find $\|u\|$, we can first find $\|u\|^2 = u \cdot u$ by squaring each component of u and adding, as follows:
$\|u\|^2 = 1^2 + (-2)^2 + (-4)^2 + 5^2 + 3^2 = 1 + 4 + 16 + 25 + 9 = 55$
Then $\|u\| = \sqrt{55}$.

We can normalize u as follows:

$$\hat{u} = \frac{u}{\|u\|} = \left(\frac{1}{\sqrt{55}}, \frac{-2}{\sqrt{55}}, \frac{-4}{\sqrt{55}}, \frac{5}{\sqrt{55}}, \frac{3}{\sqrt{55}} \right)$$

This is the unique unit vector in the same direction as u.

The following formula is known as the *Schwarz inequality* or *Cauchy-Schwarz inequality*. It is used in many branches of mathematics.

Theorem 1.3 (Schwarz): For any vectors u, v in $\mathbf{R}^n$, $|u \cdot v| \leq \|u\| \|v\|$.

The following result is known as the *triangle inequality* or *Minkowski's inequality*.

Theorem 1.4 (Minkowski): For any vectors u, v in $\mathbf{R}^n$,

$$\|u + v\| \leq \|u\| + \|v\|.$$

Chapter 2
ALGEBRA OF MATRICES

IN THIS CHAPTER:

- ✔ *Matrices*
- ✔ *Matrix Addition and Scalar Multiplication*
- ✔ *Matrix Multiplication*
- ✔ *Transpose of a Matrix*
- ✔ *Square Matrices*
- ✔ *Powers of Matrices; Polynomials in Matrices*
- ✔ *Invertible (Nonsingular) Matrices*
- ✔ *Special Types of Square Matrices*
- ✔ *Block Matrices*

Matrices

This chapter investigates matrices and algebraic operations defined on them. These matrices may be viewed as rectangular arrays of elements where each entry depends on two subscripts (as compared with

vectors, where each entry depends on only one subscript). Systems of linear equations and their solutions (Chapter 3) may be efficiently investigated using the language of matrices. The entries in our matrices will come from some arbitrary, but fixed, field K. The elements of K are called *numbers* or *scalars*. Nothing essential is lost if the reader assumes that K is the real field $\mathbf{R}$.

A *matrix A over a field K* or, simply, a *matrix A* (when K is implicit) is a rectangular array of scalars usually presented in the following form:

$$A = \begin{bmatrix} a_{11} & a_{12} & \cdots & a_{1n} \\ a_{21} & a_{22} & \cdots & a_{2n} \\ \cdots & \cdots & \cdots & \cdots \\ a_{m1} & a_{m2} & \cdots & a_{mn} \end{bmatrix}$$

The rows of such a matrix A are the m horizontal lists of scalars:

$$(a_{11}, a_{12}, \dots, a_{1n}), (a_{21}, a_{22}, \dots, a_{2n}), \dots, (a_{m1}, a_{m2}, \dots, a_{mn})$$

A matrix with m rows and n columns is called an *m by n* matrix, written $m \times n$. The pair of numbers m and n is called the *size* of the matrix. Two matrices A and B are *equal*, written $A = B$, if they have the same size and if corresponding elements are equal. Thus the equality of two $m \times n$ matrices is equivalent to a system of mn equalities, one for each corresponding pair of elements.

A matrix with only one row is called a *row matrix* or *row vector*, and a matrix with only one column is called a *column matrix* or *column vector*. A matrix whose entries are all zero is called a *zero matrix* and will usually be denoted by 0.

Example 2.1.

(*a*) The rectangular array $A = \begin{bmatrix} 1 & -4 & 5 \\ 0 & 3 & -2 \end{bmatrix}$ is a 2×3 matrix. Its rows are $(1, -4, 5)$ and $(0, 3, -2)$, and its columns are

$$\begin{bmatrix} 1 \\ 0 \end{bmatrix}, \begin{bmatrix} -4 \\ 3 \end{bmatrix}, \begin{bmatrix} 5 \\ -2 \end{bmatrix}$$

(*b*) The 2×4 zero matrix is the matrix $0 = \begin{bmatrix} 0 & 0 & 0 & 0 \\ 0 & 0 & 0 & 0 \end{bmatrix}$.

(*c*) Find x, y, z, t such that $\begin{bmatrix} x+y & 2z+t \\ x-y & z-t \end{bmatrix} = \begin{bmatrix} 3 & 7 \\ 1 & 5 \end{bmatrix}$

By definition of equality of matrices, the four corresponding entries must be equal. Thus:

$$x+y=3 \quad x-y=1 \quad 2z+t=7 \quad z-t=5$$

Solving the above system of equations yields

$$x=2,\ y=1,\ z=4,\ t=-1.$$

Matrix Addition and Scalar Multiplication

Let $A=[a_{ij}]$ and $B=[b_{ij}]$ be two matrices with the same size, say $m \times n$ matrices. The *sum* of A and B, written $A + B$, is the matrix obtained by adding corresponding elements from A and B. That is,

$$A+B=\begin{bmatrix} a_{11}+b_{11} & a_{12}+b_{12} & \dots & a_{1n}+b_{1n} \\ a_{21}+b_{21} & a_{22}+b_{22} & \dots & a_{2n}+b_{2n} \\ \dots & \dots & \dots & \dots \\ a_{m1}+b_{m1} & a_{m2}+b_{m2} & \dots & a_{mn}+b_{mn} \end{bmatrix}$$

The *product* of the matrix A by a scalar k, written $k \cdot A$ or simply kA, is the matrix obtained by multiplying each element of A by k. That is,

$$kA=\begin{bmatrix} ka_{11} & ka_{12} & \dots & ka_{1n} \\ ka_{21} & ka_{22} & \dots & ka_{2n} \\ \dots & \dots & \dots & \dots \\ ka_{m1} & ka_{m2} & \dots & ka_{mn} \end{bmatrix}$$

Observe that $A + B$ and kA are also $m \times n$ matrices.

We also define $-A = (-1)A$ and $A - B = A + (-B)$. The matrix $-A$ is called the *negative* of the matrix A, and the matrix $A - B$ is called the *difference* of A and B. The sum of matrices with different sizes is not defined.

Example 2.2. Let $A=\begin{bmatrix}1 & -2 & 3\\ 0 & 4 & 5\end{bmatrix}$ and $B=\begin{bmatrix}4 & 6 & 8\\ 1 & -3 & -7\end{bmatrix}$. Then

$$A+B=\begin{bmatrix}1+4 & -2+6 & 3+8\\ 0+1 & 4+(-3) & 5+(-7)\end{bmatrix}=\begin{bmatrix}5 & 4 & 11\\ 1 & 1 & -2\end{bmatrix}$$

$$3A=\begin{bmatrix}3(1) & 3(-2) & 3(3)\\ 3(0) & 3(4) & 3(5)\end{bmatrix}=\begin{bmatrix}3 & -6 & 9\\ 0 & 12 & 15\end{bmatrix}$$

$$2A-3B=\begin{bmatrix}2 & -4 & 6\\ 0 & 8 & 10\end{bmatrix}+\begin{bmatrix}-12 & -18 & -24\\ -3 & 9 & 21\end{bmatrix}=\begin{bmatrix}-10 & -22 & -18\\ -3 & 17 & 31\end{bmatrix}$$

The matrix $2A - 3B$ is called a *linear combination* of A and B.

Basic properties of matrices under the operations of matrix addition and scalar multiplication follow.

Theorem 2.1: Consider any matrices A, B, C (with the same size) and any scalars k and k'. Then:

(i) $(A + B) + C = A + (B + C)$,
(ii) $A + 0 = 0 + A = A$,
(iii) $A + (-A) = (-A) + A = 0$,
(iv) $A + B = B + A$,
(v) $k(A + B) = kA + kB$,
(vi) $(k + k')A = kA + k'A$,
(vii) $(kk')A = k(k'A)$,
(viii) $1 \cdot A = A$.

Note first that the 0 in (ii) and (iii) refers to the zero matrix. Also, by (i) and (iv), any sum of matrices $A_1 + A_2 + \ldots + A_n$ requires no parentheses, and the sum does not depend on the order of the matrices.

Observe the similarity between Theorem 2.1 for matrices and Theorem 1.1 for vectors. In fact, the above operations for matrices may be viewed as generalizations of the corresponding operations for vectors.

Matrix Multiplication

Before we define matrix multiplication, it will be instructive to first introduce the *summation symbol* Σ (the Greek capital letter sigma).

Suppose $f(k)$ is an algebraic expression involving the letter k. Then the expression $\sum_{k=1}^{n} f(k)$ has the following meaning. First we set $k = 1$ in

$f(k)$, obtaining $f(1)$. Then we set $k = 2$ in $f(k)$, obtaining $f(2)$, and add this to $f(1)$, obtaining $f(1) + f(2)$. Then we set $k = 3$ in $f(k)$, obtaining $f(3)$, and add this to the previous sum, obtaining $f(1) + f(2) + f(3)$. We continue this process until we obtain the sum $f(1) + f(2) + \ldots + f(n)$. Observe that at each step we increase the value of k by 1 until we reach n. The letter k is called the *index*, and 1 and n are called, respectively, the *lower* and *upper* limits. Other letters frequently used as indices are i and j.

We also generalize our definition by allowing the sum to range from any integer n_1 to any integer n_2. That is, we define

$$\sum_{k=n_1}^{n_2} f(k) = f(n_1) + f(n_1+1) + f(n_1+2) + \ldots + f(n_2)$$

Example 2.3.

$$(a)\ \sum_{k=1}^{5} x_k = x_1 + x_2 + x_3 + x_4 + x_5 \ \text{ and } \ \sum_{i=1}^{n} a_i b_i = a_1 b_1 + a_2 b_2 + \ldots + a_n b_n$$

$$(b)\ \sum_{j=2}^{5} j^2 = 2^2 + 3^2 + 4^2 + 5^2 = 54 \ \text{ and } \ \sum_{i=0}^{n} a_i x^i = a_0 + a_1 x + a_2 x^2 + \ldots + a_n x^n$$

The product of matrices A and B, written AB, is somewhat complicated. For this reason, we first begin with a special case.

The product AB of a row matrix $A = [a_i]$ and a column matrix $B = [b_i]$ with the same number of elements is defined to be the scalar (or 1×1 matrix) obtained by multiplying corresponding entries and adding; that is,

$$AB = [a_1, a_2, \ldots, a_n] \begin{bmatrix} b_1 \\ b_2 \\ \ldots \\ b_n \end{bmatrix} = a_1 b_1 + a_2 b_2 + \ldots + a_n b_n = \sum_{k=1}^{n} a_k b_k$$

We emphasize that AB is a scalar (or a 1×1 matrix). The product AB is not defined when A and B have different numbers of elements.

Example 2.4.

$$(a)\quad [7,-4,5]\begin{bmatrix} 3 \\ 2 \\ -1 \end{bmatrix} = 7(3)+(-4)(2)+5(-1) = 21-8-5 = 8$$

$$(b)\quad [6,-1,8,3]\begin{bmatrix} 4 \\ -9 \\ -2 \\ 5 \end{bmatrix} = 24+9-16+15 = 32$$

We are now ready to define matrix multiplication in general. Suppose $A = [a_{jk}]$ and $B = [b_{kj}]$ are matrices such that the number of columns of A is equal to the number of rows of B; say, A is an $m \times p$ matrix and B is a $p \times n$ matrix. Then the product AB is the $m \times n$ matrix whose ij-entry is obtained by multiplying the ith row of A by the jth column of B. That is,

$$\begin{bmatrix} a_{11} & \dots & a_{1p} \\ \cdot & \dots & \cdot \\ a_{i1} & \dots & a_{ip} \\ \cdot & \dots & \cdot \\ a_{m1} & \dots & a_{mp} \end{bmatrix} \begin{bmatrix} b_{11} & \dots & b_{1j} & \dots & b_{1n} \\ \cdot & \dots & \cdot & \dots & \cdot \\ \cdot & \dots & \cdot & \dots & \cdot \\ \cdot & \dots & \cdot & \dots & \cdot \\ b_{p1} & \dots & b_{pj} & \dots & b_{pn} \end{bmatrix} = \begin{bmatrix} c_{11} & \dots & c_{1n} \\ \cdot & \dots & \cdot \\ \cdot & c_{ij} & \cdot \\ \cdot & \dots & \cdot \\ c_{m1} & \dots & c_{mn} \end{bmatrix}$$

where $c_{ij} = a_{i1}b_{1j} + a_{i2}b_{2j} + \ldots + a_{ip}b_{pj} = \sum_{k=1}^{p} a_{ik}b_{kj}$.

The product AB is not defined if A is an $m \times p$ matrix and B is a $q \times n$ matrix, where $p \neq q$.

Example 2.5.

(*a*) Find AB where $A = \begin{bmatrix} 1 & 3 \\ 2 & -1 \end{bmatrix}$ and $B = \begin{bmatrix} 2 & 0 & -4 \\ 5 & -2 & 6 \end{bmatrix}$.

Since A is 2×2 and B is 2×3, the product AB is defined and AB is a 2×3 matrix. To obtain the first row of the product matrix AB, multiply the first row [1, 3] of A by each column of B,

$$\begin{bmatrix} 2 \\ 5 \end{bmatrix}, \begin{bmatrix} 0 \\ -2 \end{bmatrix}, \begin{bmatrix} -4 \\ 6 \end{bmatrix}$$

respectively. That is,

$$AB = \begin{bmatrix} 2+15 & 0-6 & -4+18 \\ & & \end{bmatrix} = \begin{bmatrix} 17 & -6 & 14 \\ & & \end{bmatrix}$$

To obtain the second row of AB, multiply the second row $[2, -1]$ of A by each column of B. Thus

$$AB = \begin{bmatrix} 17 & -6 & 14 \\ 4-5 & 0+2 & -8-6 \end{bmatrix} = \begin{bmatrix} 17 & -6 & 14 \\ -1 & 2 & -14 \end{bmatrix}$$

(*b*) Suppose $A = \begin{bmatrix} 1 & 2 \\ 3 & 4 \end{bmatrix}$ and $B = \begin{bmatrix} 5 & 6 \\ 0 & -2 \end{bmatrix}$. Then

$$AB = \begin{bmatrix} 5+0 & 6-4 \\ 15+0 & 18-8 \end{bmatrix} = \begin{bmatrix} 5 & 2 \\ 15 & 10 \end{bmatrix} \text{ and}$$

$$BA = \begin{bmatrix} 5+18 & 10+24 \\ 0-6 & 0-8 \end{bmatrix} = \begin{bmatrix} 23 & 34 \\ -6 & -8 \end{bmatrix}$$

The above example shows that matrix multiplication is not commutative, i.e., the products AB and BA of matrices need not be equal. However, matrix multiplication does satisfy the following properties.

Theorem 2.2: Let A, B, C be matrices. Then, whenever the products and sums are defined:

(i) $(AB)C = A(BC)$ (associative law),
(ii) $A(B + C) = AB + AC$ (left distributive law),
(iii) $(B + C)A = BA + CA$ (right distributive law),
(iv) $k(AB) = (kA)B = A(kB)$, where k is a scalar.

We note that $0A = 0$ and $B0 = 0$, where 0 is the zero matrix.

Transpose of a Matrix

The *transpose* of a matrix A, written A^T, is the matrix obtained by writing the columns of A, in order, as rows. For example,

$$\begin{bmatrix} 1 & 2 & 3 \\ 4 & 5 & 6 \end{bmatrix}^T = \begin{bmatrix} 1 & 4 \\ 2 & 5 \\ 3 & 6 \end{bmatrix} \text{ and } [1,-3,-5]^T = \begin{bmatrix} 1 \\ -3 \\ -5 \end{bmatrix}$$

In other words, if $A = [a_{ij}]$ is an $m \times n$ matrix, then $A^T = [b_{ij}]$ is the $n \times m$ matrix where $b_{ij} = a_{ji}$.

Observe that the transpose of a row vector is a column vector. Similarly, the transpose of a column vector is a row vector.

The next theorem lists basic properties of the transpose operation.

Theorem 2.3: Let A and B be matrices and let k be a scalar. Then, whenever the sum and product are defined:

(i) $(A + B)^T = A^T + B^T$, (iii) $(kA)^T = kA^T$,

(ii) $(A^T)^T = A$, (iv) $(AB)^T = B^T A^T$.

We emphasize that, by (iv), the transpose of a product is the product of the transposes, but in the reverse order.

Square Matrices

A *square matrix* is a matrix with the same number of rows as columns. An $n \times n$ square matrix is said to be of *order n* and is sometimes called an *n-square matrix*.

Recall that not every two matrices can be added or multiplied. However, if we only consider square matrices of some given order n, then this inconvenience disappears. Specifically, the operations of addition, multiplication, scalar multiplication, and transpose can be performed on any $n \times n$ matrices, and the result is again an $n \times n$ matrix.

Example 2.6. The following are square matrices of order 3:

$$A = \begin{bmatrix} 1 & 2 & 3 \\ -4 & -4 & -4 \\ 5 & 6 & 7 \end{bmatrix} \text{ and } B = \begin{bmatrix} 2 & -5 & 1 \\ 0 & 3 & -2 \\ 1 & 2 & -4 \end{bmatrix}$$

The following are also matrices of order 3:

$$A+B=\begin{bmatrix} 3 & -3 & 4 \\ -4 & -1 & -6 \\ 6 & 8 & 3 \end{bmatrix} \quad 2A=\begin{bmatrix} 2 & 4 & 6 \\ -8 & -8 & -8 \\ 10 & 12 & 14 \end{bmatrix} \quad A^T=\begin{bmatrix} 1 & -4 & 5 \\ 2 & -4 & 6 \\ 3 & -4 & 7 \end{bmatrix}$$

$$AB=\begin{bmatrix} 5 & 7 & -15 \\ -12 & 0 & 20 \\ 17 & 7 & -35 \end{bmatrix} \quad BA=\begin{bmatrix} 27 & 30 & 33 \\ -22 & -24 & -26 \\ -27 & -30 & -33 \end{bmatrix}$$

Diagonal and Trace

Let $A = [a_{ij}]$ be an n-square matrix. The *diagonal* or *main diagonal* of A consists of the elements with the same subscripts, that is,

$$a_{11}, a_{22}, a_{33}, \ldots, a_{nn}.$$

The *trace* of A, written $\mathrm{tr}(A)$, is the sum of the diagonal elements. Namely, $\mathrm{tr}(A) = a_{11} + a_{22} + a_{33} + \ldots + a_{nn}.$

The following theorem applies.

Theorem 2.4: Suppose $A = [aij]$ and $B = [bij]$ are n-square matrices and k is a scalar. Then:

(i) $\mathrm{tr}(A + B) = \mathrm{tr}(A) + \mathrm{tr}(B)$,
(ii) $\mathrm{tr}(kA) = k\,\mathrm{tr}(A)$,
(iii) $\mathrm{tr}(A^T) = \mathrm{tr}(A)$
(iv) $\mathrm{tr}(AB) = \mathrm{tr}(BA)$.

Example 2.7. Let A and B be the matrices A and B in Example 2.6. Then

diagonal of $A = [1, -4, 7]$ and $\mathrm{tr}(A) = 1 - 4 + 7 = 4$
diagonal of $B = [2, 3, -4]$ and $\mathrm{tr}(B) = 2 + 3 - 4 = 1$

Moreover,

$$\mathrm{tr}(A + B) = 3 - 1 + 3 = 5, \quad \mathrm{tr}(2A) = 2 - 8 + 14 = 8,$$
$$\mathrm{tr}(A^T) = 1 - 4 + 7 = 4$$
$$\mathrm{tr}(AB) = 5 + 0 - 35 = -30, \quad \mathrm{tr}(BA) = 27 - 24 - 33 = -30$$

As expected from Theorem 2.4,

$$\mathrm{tr}(A+B) = \mathrm{tr}(A) + \mathrm{tr}(B), \quad \mathrm{tr}(A^T) = \mathrm{tr}(A), \quad \mathrm{tr}(2A) = 2\mathrm{tr}(A)$$

Furthermore, although $AB \neq BA$, the traces are equal.

Identity Matrix, Scalar Matrices

The n-square *identity* or *unit* matrix, denoted by I_n, or simply I, is the n-square matrix with 1s on the diagonal and 0s elsewhere. The identity matrix I is similar to the scalar 1 in that, for any n-square matrix A, $AI = IA = A$. More generally, if B is an $m \times n$ matrix, then

$$BI_n = I_m B = B.$$

For any scalar k, the matrix kI that contains k's on the diagonal and 0's elsewhere is called the scalar matrix corresponding to the scalar k. Observe that $(kI)A = k(IA) = kA$. That is, multiplying a matrix A by the scalar matrix kI is equivalent to multiplying A by the scalar k.

Example 2.8. The following are the identity matrices of orders 3 and 4 and the corresponding scalar matrices for $k = 5$:

$$\begin{bmatrix} 1 & 0 & 0 \\ 0 & 1 & 0 \\ 0 & 0 & 1 \end{bmatrix}, \quad \begin{bmatrix} 1 & & & \\ & 1 & & \\ & & 1 & \\ & & & 1 \end{bmatrix}, \quad \begin{bmatrix} 5 & 0 & 0 \\ 0 & 5 & 0 \\ 0 & 0 & 5 \end{bmatrix}, \quad \begin{bmatrix} 5 & & & \\ & 5 & & \\ & & 5 & \\ & & & 5 \end{bmatrix}$$

Note ✓

It is common practice to omit blocks or patterns of 0's when there is no ambiguity, as in the above second and fourth matrices.

The *Kronecker delta* function δ_{ij} is defined by

$$\delta_{ij} = \begin{cases} 0 & \text{if} \quad i \neq j \\ 1 & \text{if} \quad i = j \end{cases}$$

Thus the identity matrix may be defined by $I = [\delta_{ij}]$.

Powers of Matrices; Polynomials in Matrices

Let A be an n-square matrix over a field K. *Powers* of A are defined as follows:

$$A^2 = AA, \quad A^3 = A^2A, \quad \ldots, \quad A^{n+1} = A^nA, \quad \ldots, \quad \text{and} \quad A^0 = I$$

Polynomials in the matrix A are also defined. Specifically, for any polynomial $f(x) = a_0 + a_1x + a_2x^2 + \ldots + a_nx^n$ where the a_i are scalars in K, $f(A)$ is defined to be the following matrix:

$$f(A) = a_0I + a_1A + a_2A^2 + \ldots + a_nA^n$$

If $f(A)$ is the zero matrix, then A is called a *zero* or *root* of $f(x)$.

Example 2.9. Suppose $A = \begin{bmatrix} 1 & 2 \\ 3 & -4 \end{bmatrix}$. Then

$$A^2 = \begin{bmatrix} 1 & 2 \\ 3 & -4 \end{bmatrix}\begin{bmatrix} 1 & 2 \\ 3 & -4 \end{bmatrix} = \begin{bmatrix} 7 & -6 \\ -9 & 22 \end{bmatrix}$$

and

$$A^3 = A^2A = \begin{bmatrix} 7 & -6 \\ -9 & 22 \end{bmatrix}\begin{bmatrix} 1 & 2 \\ 3 & -4 \end{bmatrix} = \begin{bmatrix} -11 & 38 \\ 57 & -106 \end{bmatrix}$$

Suppose $f(x) = 2x^2 - 3x + 5$ and $g(x) = x^2 + 3x - 10$. Then

$$f(A) = 2\begin{bmatrix} 7 & -6 \\ -9 & 22 \end{bmatrix} - 3\begin{bmatrix} 1 & 2 \\ 3 & -4 \end{bmatrix} + 5\begin{bmatrix} 1 & 0 \\ 0 & 1 \end{bmatrix} = \begin{bmatrix} 16 & -18 \\ -27 & 61 \end{bmatrix}$$

$$g(A) = \begin{bmatrix} 7 & -6 \\ -9 & 22 \end{bmatrix} + 3\begin{bmatrix} 1 & 2 \\ 3 & -4 \end{bmatrix} - 10\begin{bmatrix} 1 & 0 \\ 0 & 1 \end{bmatrix} = \begin{bmatrix} 0 & 0 \\ 0 & 0 \end{bmatrix}$$

Thus A is a zero of the polynomial $g(x)$.

Invertible (Nonsingular) Matrices

A square matrix A is said to be *invertible* or *nonsingular* if there exists a matrix B such that $AB = BA = I$ where I is the identity matrix. Such a matrix B is unique. That is, if $AB_1 = B_1A = I$ and $AB_2 = B_2A = I$, then $B_1 = B_1I = B_1(AB_2) = (B_1A)B_2 = IB_2 = B_2$.

We call such a matrix B the *inverse* of A and denote it by A^{-1}. Observe that the above relation is symmetric; that is, if B is the inverse of A, then A is the inverse of B.

Example 2.10. Suppose that $A = \begin{bmatrix} 2 & 5 \\ 1 & 3 \end{bmatrix}$ and $B = \begin{bmatrix} 3 & -5 \\ -1 & 2 \end{bmatrix}$. Then

$$AB = \begin{bmatrix} 6-5 & -10+10 \\ 3-3 & -5+6 \end{bmatrix} = \begin{bmatrix} 1 & 0 \\ 0 & 1 \end{bmatrix} \text{ and } BA = \begin{bmatrix} 6-5 & 15-15 \\ -2+2 & -5+6 \end{bmatrix} = \begin{bmatrix} 1 & 0 \\ 0 & 1 \end{bmatrix}$$

Thus A and B are inverses.

It is known that $AB = I$ if and only if $BA = I$. Thus it is necessary to test only one product to determine whether or not two given matrices are inverses.

Now suppose A and B are invertible. Then AB is invertible and $(AB)^{-1} = B^{-1}A^{-1}$. More generally, if $A_1, A_2, \ldots, A_k$ are invertible, then their product is invertible and $(A_1A_2 \ldots A_k)^{-1} = A_k^{-1} \ldots A_2^{-1}A_1^{-1}$, the product of the inverses in the reverse order.

Inverse of a 2 × 2 Matrix

Let A be an arbitrary 2×2 matrix, say $A = \begin{bmatrix} a & b \\ c & d \end{bmatrix}$. We want to derive a formula for A^{-1}, the inverse of A. Specifically, we seek $2^2 = 4$ scalars, say x_1, y_1, x_2, y_2, such that

$$\begin{bmatrix} a & b \\ c & d \end{bmatrix}\begin{bmatrix} x_1 & x_2 \\ y_1 & y_2 \end{bmatrix} = \begin{bmatrix} 1 & 0 \\ 0 & 1 \end{bmatrix} \text{ or } \begin{bmatrix} ax_1 + by_1 & ax_2 + by_2 \\ cx_1 + dy_1 & cx_2 + dy_2 \end{bmatrix} = \begin{bmatrix} 1 & 0 \\ 0 & 1 \end{bmatrix}$$

Setting the four entries equal to the corresponding entries in the identity matrix yields four equations, which can be partitioned into two 2×2 systems as follows:

$$ax_1 + by_1 = 1, \quad ax_2 + by_2 = 0$$
$$cx_1 + dy_1 = 0, \quad cx_2 + dy_2 = 1$$

Suppose we let $|A| = ad - bc$ (called the *determinant* of A). Assuming $|A| \neq 0$, we can solve uniquely for the above unknowns x_1, y_1, x_2, y_2, obtaining

$$x_1 = \frac{d}{|A|} \quad y_1 = \frac{-c}{|A|} \quad x_2 = \frac{-b}{|A|} \quad y_2 = \frac{a}{|A|}$$

Accordingly,

$$A^{-1} = \begin{bmatrix} a & b \\ c & d \end{bmatrix}^{-1} = \begin{bmatrix} d/|A| & -b/|A| \\ -c/|A| & a/|A| \end{bmatrix} = \frac{1}{|A|}\begin{bmatrix} d & -b \\ -c & a \end{bmatrix}$$

In other words, when $|A| \neq 0$, the inverse of a 2×2 matrix A may be obtained from A as follows:

(1) Interchange the two elements on the diagonal.
(2) Take the negatives of the other two elements.
(3) Multiply the resulting matrix by $1/|A|$ or, equivalently, divide each element by $|A|$.

In case $|A| = 0$, the matrix A is not invertible.

Example 2.11. Find the inverse of $A = \begin{bmatrix} 2 & 3 \\ 4 & 5 \end{bmatrix}$ and $B = \begin{bmatrix} 1 & 3 \\ 2 & 6 \end{bmatrix}$.

First evaluate $|A| = 2(5) - 3(4) = 10 - 12 = -2$. Since $|A| \neq 0$, the matrix A is invertible and $A^{-1} = \frac{1}{-2}\begin{bmatrix} 5 & -3 \\ -4 & 2 \end{bmatrix} = \begin{bmatrix} \frac{-5}{2} & \frac{3}{2} \\ 2 & -1 \end{bmatrix}$.

Now evaluate $|B| = 1(6) - 3(2) = 0$. Since $|B| = 0$, the matrix B has no inverse.

REMEMBER

The above property that a matrix is invertible if and only if A has a nonzero determinant is true for square matrices of any order.

Suppose A is an arbitrary n-square matrix. Finding its inverse A^{-1} reduces, as above, to finding the solution of a collection of $n \times n$ systems of linear equations. The solution of such systems and an efficient way of solving such a collection of systems is treated in Chapter 3.

Special Types of Square Matrices

This section describes a number of special kinds of square matrices.

Diagonal Matrices

A square matrix $D = [d_{ij}]$ is *diagonal* if its nondiagonal entries are all zero. Such a matrix is sometimes denoted by

$$D = \operatorname{diag}(d_{11}, d_{22}, \ldots, d_{nn})$$

where some or all the d_{ii} may be zero. For example,

$$\begin{bmatrix} 4 & 0 \\ 0 & -5 \end{bmatrix}, \quad \begin{bmatrix} 3 & & \\ & -7 & \\ & & 0 \end{bmatrix}$$

are diagonal matrices, which may be represented, respectively, by diag(4, –5), and diag(3, –7, 0). (The patterns of 0's in the second matrix have been omitted.)

Triangular Matrices

A square matrix is *upper triangular* or simply *triangular* if all entries below the (main) diagonal are equal to 0, that is, if $a_{ij} = 0$ for all $i > j$. Generic upper triangular matrices of orders 2 and 3 are as follows:

$$\begin{bmatrix} a_{11} & a_{12} \\ 0 & a_{22} \end{bmatrix} \quad \begin{bmatrix} b_{11} & b_{12} & b_{13} \\ & b_{22} & b_{23} \\ & & b_{33} \end{bmatrix}$$

(As with diagonal matrices, it is common to omit patterns of 0's.)

Theorem 2.5: Suppose $A = [a_{ij}]$ and $B = [b_{ij}]$ are $n \times n$ (upper) triangular matrices. Then:

(i) $A + B$, kA, AB are triangular with respective diagonals.
$(a_{11} + b_{11}, \ldots, a_{nn} + b_{nn})$, $(ka_{11}, \ldots, ka_{nn})$, $(a_{11}b_{11}, \ldots, a_{nn}b_{nn})$,

(ii) For any polynomial $f(x)$, the matrix $f(A)$ is triangular with diagonal $(f(a_{11}), f(a_{22}), \ldots, f(a_{nn}))$

(iii) A is invertible if and only if each diagonal element $a_{ii} \neq 0$, and when A^{-1} exists it is also triangular.

A *lower triangular* matrix is a square matrix whose entries above the diagonal are all zero. We note that Theorem 2.5 is true if we replace "triangular" by either "lower triangular" or "diagonal."

Symmetric Matrices

A matrix A is *symmetric* if $A^T = A$. Equivalently, $A = [a_{ij}]$ is symmetric if *symmetric elements* (mirror elements with respect to the diagonal) are equal, that is, if each $a_{ij} = a_{ji}$.

A matrix A is *skew-symmetric* if $A^T = -A$ or, equivalently, if each $a_{ij} = -a_{ji}$. Clearly the diagonal elements of such a matrix must be zero, since $a_{ii} = -a_{ii}$ implies $a_{ii} = 0$.

(Note that a matrix A must be square if $A^T = A$ or $A^T = -A$.)

Example 2.12. Let $A = \begin{bmatrix} 2 & -3 & 5 \\ -3 & 6 & 7 \\ 5 & 7 & -8 \end{bmatrix}$, $B = \begin{bmatrix} 0 & 3 & -4 \\ -3 & 0 & 5 \\ 4 & -5 & 0 \end{bmatrix}$.

By inspection, the symmetric elements in A are equal, or $A^T = A$. Thus A is symmetric.

The diagonal elements of B are 0 and symmetric elements are negatives of each other, or $B^T = -B$. Thus B is skew-symmetric.

Orthogonal Matrices

A real matrix A is *orthogonal* if $A^T = A^{-1}$, that is, if $AA^T = A^TA = I$. Thus A must necessarily be square and invertible.

Example 2.13. Let $A = \begin{bmatrix} \frac{1}{9} & \frac{8}{9} & -\frac{4}{9} \\ \frac{4}{9} & -\frac{4}{9} & -\frac{7}{9} \\ \frac{8}{9} & \frac{1}{9} & \frac{4}{9} \end{bmatrix}$.

$$AA^T = \begin{bmatrix} \frac{1}{9} & \frac{8}{9} & -\frac{4}{9} \\ \frac{4}{9} & -\frac{4}{9} & -\frac{7}{9} \\ \frac{8}{9} & \frac{1}{9} & \frac{4}{9} \end{bmatrix} \begin{bmatrix} \frac{1}{9} & \frac{4}{9} & \frac{8}{9} \\ \frac{8}{9} & -\frac{4}{9} & \frac{1}{9} \\ -\frac{4}{9} & -\frac{7}{9} & \frac{4}{9} \end{bmatrix} = \begin{bmatrix} 1 & 0 & 0 \\ 0 & 1 & 0 \\ 0 & 0 & 1 \end{bmatrix}$$

Multiplying A by A^T yields I; that is $AA^T = I$. This means $A^TA = I$, as well. Thus $A^T = A^{-1}$; that is, A is orthogonal.

Now suppose A is a real orthogonal 3×3 matrix with rows

$$u_1 = (a_1, a_2, a_3),\ u_2 = (b_1, b_2, b_3),\ u_3 = (c_1, c_2, c_3).$$

Since A is orthogonal, we must have $AA^T = I$. Namely

$$AA^T = \begin{bmatrix} a_1 & a_2 & a_3 \\ b_1 & b_2 & b_3 \\ c_1 & c_2 & c_3 \end{bmatrix}\begin{bmatrix} a_1 & b_1 & c_1 \\ a_2 & b_2 & c_2 \\ a_3 & b_3 & c_3 \end{bmatrix} = \begin{bmatrix} 1 & 0 & 0 \\ 0 & 1 & 0 \\ 0 & 0 & 1 \end{bmatrix} = I$$

Multiplying A by A^T and setting each entry equal to the corresponding entry in I yields the following nine equations:

$$\begin{aligned} a_1^2 + a_2^2 + a_3^2 = 1, \quad & a_1b_1 + a_2b_2 + a_3b_3 = 0, \quad & a_1c_1 + a_2c_2 + a_3c_3 = 0 \\ b_1a_1 + b_2a_2 + b_3a_3 = 0, \quad & b_1^2 + b_2^2 + b_3^2 = 1, \quad & b_1c_1 + b_2c_2 + b_3c_3 = 0 \\ c_1a_1 + c_2a_2 + c_3a_3 = 0, \quad & c_1b_1 + c_2b_2 + c_3b_3 = 0, \quad & c_1^2 + c_2^2 + c_3^2 = 1 \end{aligned}$$

Accordingly, $u_1 \cdot u_1 = 1$, $u_2 \cdot u_2 = 1$, $u_3 \cdot u_3 = 1$, and $u_i \cdot u_j = 0$ for $i \neq j$. Thus the rows u_1, u_2, u_3 are unit vectors and are orthogonal to each other.

Generally speaking, vectors $u_1, u_2, \ldots, u_m$ in $\mathbf{R}^n$ are said to form an *orthonormal set* of vectors if the vectors are unit vectors and are orthogonal to each other, that is,

$$u_i \cdot u_j = \begin{cases} 0 & \text{if} \quad i \neq j \\ 1 & \text{if} \quad i = j \end{cases}$$

In other words, $u_i \cdot u_j = \delta_{ij}$ where δ_{ij} is the Kronecker delta function.

We have shown that the condition $AA^T = I$ implies that the rows of A form an orthonormal set of vectors. The condition $A^TA = I$ similarly implies that the columns of A also form an orthonormal set of vectors. Furthermore, since each step is reversible, the converse is true.

The above results for 3×3 matrices is true in general. That is, the following theorem holds.

Theorem 2.6: Let A be a real matrix. Then the following are equivalent:

(i) A is orthogonal.
(ii) The rows of A form an orthonormal set.
(iii) The columns of A form an orthonormal set.

Normal vectors

A real matrix A is *normal* if it *commutes* with its transpose A^T, that is, if $AA^T = A^TA$. If A is symmetric, orthogonal, or skew-symmetric, then A is normal. There are also other normal matrices.

Example 2.14. Let $A = \begin{bmatrix} 6 & -3 \\ 3 & 6 \end{bmatrix}$. Then

$$AA^T = \begin{bmatrix} 6 & -3 \\ 3 & 6 \end{bmatrix} \begin{bmatrix} 6 & 3 \\ -3 & 6 \end{bmatrix} = \begin{bmatrix} 45 & 0 \\ 0 & 45 \end{bmatrix}$$

and

$$A^TA = \begin{bmatrix} 6 & 3 \\ -3 & 6 \end{bmatrix} \begin{bmatrix} 6 & -3 \\ 3 & 6 \end{bmatrix} = \begin{bmatrix} 45 & 0 \\ 0 & 45 \end{bmatrix}$$

Since $AA^T = A^TA$, the matrix A is normal.

Block Matrices

Using a system of horizontal and vertical (dashed) lines, we can partition a matrix A into submatrices called *blocks* (or *cells*) of A. Clearly a given matrix may be divided into blocks in different ways. For example,

$$\left[\begin{array}{cc|cc|c} 1 & -2 & 0 & 1 & 3 \\ 2 & 3 & 5 & 7 & -2 \\ \hline 3 & 1 & 4 & 5 & 9 \\ 4 & 6 & -3 & 1 & 8 \end{array}\right] \quad \left[\begin{array}{cc|ccc} 1 & -2 & 0 & 1 & 3 \\ \hline 2 & 3 & 5 & 7 & -2 \\ 3 & 1 & 4 & 5 & 9 \\ \hline 4 & 6 & -3 & 1 & 8 \end{array}\right]$$

$$\left[\begin{array}{ccc|cc} 1 & -2 & 0 & 1 & 3 \\ 2 & 3 & 5 & 7 & -2 \\ \hline 3 & 1 & 4 & 5 & 9 \\ 4 & 6 & -3 & 1 & 8 \end{array}\right]$$

The convenience of the partition of matrices, say A and B, into blocks is that the result of operations on A and B can be obtained by carrying out the computation with the blocks, just as if they were the actual elements of the matrices. This is illustrated below, where the notation $A = [A_{ij}]$ will be used for a block matrix A with blocks A_{ij}.

Suppose that $A = [A_{ij}]$ and $B = [B_{ij}]$ are block matrices with the same numbers of row and column blocks, and suppose that corresponding blocks have the same size. Then adding the corresponding blocks of A and B also adds the corresponding elements of A and B and multiplying each block of A by a scalar k multiplies each element of A by k. Thus

$$A+B=\begin{bmatrix} A_{11}+B_{11} & A_{12}+B_{12} & \dots & A_{1n}+B_{1n} \\ A_{21}+B_{21} & A_{22}+B_{22} & \dots & A_{2n}+B_{2n} \\ \dots & \dots & \dots & \dots \\ A_{m1}+B_{m1} & A_{m2}+B_{m2} & \dots & A_{mn}+B_{mn} \end{bmatrix}$$

and

$$kA=\begin{bmatrix} kA_{11} & kA_{12} & \dots & kA_{1n} \\ kA_{21} & kA_{22} & \dots & kA_{2n} \\ \dots & \dots & \dots & \dots \\ kA_{m1} & kA_{m2} & \dots & kA_{mn} \end{bmatrix}$$

The case of matrix multiplication is less obvious, but still true. That is, suppose that $U=[U_{ik}]$ and $V=[V_{kj}]$ are block matrices such that the number of columns of each block U_{ik} is equal to the number of rows of each block V_{kj}. (Thus each product $U_{ik}V_{kj}$ is defined.) Then

$$UV=\begin{bmatrix} W_{11} & W_{12} & \dots & W_{1n} \\ W_{21} & W_{22} & \dots & W_{2n} \\ \dots & \dots & \dots & \dots \\ W_{m1} & W_{m2} & \dots & W_{mn} \end{bmatrix},$$

where $W_{ij}=U_{i1}V_{1j}+U_{i2}V_{2j}+\ldots+U_{ip}V_{pj}$

Square Block Matrices

Let M be a block matrix. Then M is called a *square block matrix* if:

(i) M is a square matrix.
(ii) The blocks form a square matrix.
(iii) The diagonal blocks are also square matrices.

The latter two conditions will occur if and only if there are the same number of horizontal and vertical lines and they are placed symmetrically.

Consider the following two block matrices:

$$A = \left[\begin{array}{cc|cc|c} 1 & 2 & 3 & 4 & 5 \\ 1 & 1 & 1 & 1 & 1 \\ \hline 9 & 8 & 7 & 6 & 5 \\ \hline 4 & 4 & 4 & 4 & 4 \\ 3 & 5 & 3 & 5 & 3 \end{array}\right] \text{ and } B = \left[\begin{array}{cc|cc|c} 1 & 2 & 3 & 4 & 5 \\ 1 & 1 & 1 & 1 & 1 \\ \hline 9 & 8 & 7 & 6 & 5 \\ 4 & 4 & 4 & 4 & 4 \\ \hline 3 & 5 & 3 & 5 & 3 \end{array}\right]$$

The block matrix A is not a square block matrix, since the second and third diagonal blocks are not square. On the other hand, the block matrix B is a square block matrix.

Block Diagonal Matrices

Let $M = [A_{ij}]$ be a square block matrix such that the nondiagonal blocks are all zero matrices, that is, $A_{ij} = 0$ when $i \neq j$. Then M is called a *block diagonal matrix*. We sometimes denote such a block diagonal matrix by writing

$$M = diag(A_{11}, A_{22}, \ldots, A_{rr}) \text{ or } M = A_{11} \oplus A_{22} \oplus \ldots \oplus A_{rr}$$

The importance of block diagonal matrices is that the algebra of the block matrix is frequently reduced to the algebra of the individual blocks. Specifically, suppose $f(x)$ is a polynomial and M is the above block diagonal matrix. Then $f(M)$ is a block diagonal matrix and

$$f(M) = \text{diag}(f(A_{11}), f(A_{22}), \ldots, f(A_{rr}))$$

Also, M is invertible if and only if each A_{ii} is invertible, and, in such a case, M^{-1} is a block diagonal matrix and

$$M^{-1} = diag(A_{11}^{-1}, A_{22}^{-1}, \dots, A_{rr}^{-1})$$

Example 2.15. Determine which of the following square block matrices are upper triangular, lower triangular, or diagonal.

$$A = \left[\begin{array}{cc:c} 1 & 2 & 0 \\ 3 & 4 & 5 \\ \hdashline 0 & 0 & 6 \end{array}\right], \quad B = \left[\begin{array}{c:cc:c} 1 & 0 & 0 & 0 \\ \hdashline 2 & 3 & 4 & 0 \\ 5 & 0 & 6 & 0 \\ \hdashline 0 & 7 & 8 & 9 \end{array}\right],$$

$$C = \left[\begin{array}{c:cc} 1 & 0 & 0 \\ \hdashline 0 & 2 & 3 \\ 0 & 4 & 5 \end{array}\right], \quad D = \left[\begin{array}{cc:c} 1 & 2 & 0 \\ 3 & 4 & 5 \\ \hdashline 0 & 6 & 7 \end{array}\right]$$

(*a*) A is upper triangular since the block below the diagonal is a zero block.

(*b*) B is lower triangular since all blocks above the diagonal are zero blocks.

(*c*) C is diagonal since the blocks above and below the diagonal are zero blocks.

(*d*) D is neither upper triangular nor lower triangular. Also, no other partitioning of D will make it into either a block upper triangular matrix or a block lower triangular matrix.

Chapter 3
SYSTEMS OF LINEAR EQUATIONS

IN THIS CHAPTER:

- ✔ *Basic Definitions*
- ✔ *Equivalent Systems and Elementary Operations*
- ✔ *Small Square Systems of Linear Equations*
- ✔ *Systems in Triangular and Echelon Form*
- ✔ *Gaussian Elimination*
- ✔ *Echelon Matrices; Row Canonical Form*
- ✔ *Gaussian Elimination (Matrices)*
- ✔ *Application to Systems of Linear Equations*
- ✔ *Inverse Matrices*

Basic Definitions

Systems of linear equations play an important and motivating role in the subject of linear algebra. In fact, many problems in linear algebra reduce to finding the solution of a system of linear equations. Thus the techniques introduced in this chapter will be applicable to abstract ideas introduced later. On the other hand, some of the abstract results will give us new insights into the structure and properties of systems of linear equations.

All our systems of linear equations involve scalars as both coefficients and constants, and such scalars may come from any number field K. There is almost no loss in generality if the reader assumes that all our scalars are real numbers, that is, that they come from the real field **R**.

Linear Equation and Solutions

A *linear equation* in unknowns $x_1, x_2, \dots, x_n$ is an equation that can be put in the *standard form*

$$a_1x_1 + a_2x_2 + \dots + a_nx_n = b \tag{3.1}$$

where $a_1, a_2, \dots, a_n$, and b are constants. The constant a_k is called the *coefficient* of x_k, and b is called the *constant term* of the equation.

A *solution* of the linear equation (3.1) is a list of values for the unknowns or, equivalently, a vector u in K^n, say

$$x_1 = k_1, x_2 = k_2, \dots, x_n = k_n, \text{ or } u = (k_1, k_2, \dots, k_n)$$

such that the following statement (obtained by substituting k_i for x_i in the equation) is true:

$$a_1k_1 + a_2k_2 + \dots + a_nk_n = b$$

In such a case we say that u *satisfies* the equation.

Note!

In order to avoid subscripts, we will usually use *x*, *y* for two unknowns, *x*, *y*, *z* for three unknowns, and *x*, *y*, *z*, *t* for four unknowns, and they will be ordered as shown.

Example 3.1. Consider the linear equation $x + 2y - 3z = 6$ in three unknowns x, y, z. We note that $x = 5$, $y = 2$, $z = 1$ or, equivalently, the vector $u = (5, 2, 1)$ is a solution of the equation. That is,

$$5 + 2(2) - 3(1) = 6 + 4 - 3 = 6.$$

On the other hand, $w = (1, 2, 3)$ is not a solution, since, on substitution, we do not get a true statement:

$$1 + 2(2) - 3(3) = 1 + 4 - 9 = -4 \neq 6.$$

System of Linear Equations

A *system* of linear equations is a list of linear equations with the same unknowns. In particular, a system of m linear equations $L_1, L_2, \ldots, L_m$ in n unknowns $x_1, x_2, \ldots, x_n$ can be put in the standard form

$$\begin{aligned} a_{11}x_1 + a_{12}x_2 + \ldots + a_{1n}x_n &= b_1 \\ a_{21}x_1 + a_{22}x_2 + \ldots + a_{2n}x_n &= b_2 \\ &\cdots\cdots\cdots \\ a_{m1}x_1 + a_{m2}x_2 + \ldots + a_{mn}x_n &= b_m \end{aligned} \tag{3.2}$$

where the a_{ij} and b_i are constants. The number a_{ij} is the *coefficient* of the unknown x_j in the equation L_i and the number b_i is the *constant* of the equation L_i.

The system (3.2) is called an $m \times n$ (read: m by n) system. It is called a *square system* if $m = n$, that is, if the number m of equations is equal to the number n of unknowns.

The system (3.2) is said to be *homogeneous* if all the constant terms are zeros, that is, if $b_1 = 0, b_2 = 0, \ldots, b_m = 0$. Otherwise the system is said to be *nonhomogeneous*.

A *solution* (or a *particular solution*) of the system (3.2) is a list of values for the unknowns or, equivalently, a vector u in K^n, that is a solution of each of the equations in the system. The set of all solutions of the system is called the *solution set* or the *general solution* of the system.

Example 3.2. Consider the following system of linear equations:

$$\begin{aligned} x_1 + x_2 + 4x_3 + 3x_4 &= 5 \\ 2x_1 + 3x_2 + x_3 - 2x_4 &= 1 \\ x_1 + 2x_2 - 5x_3 + 4x_4 &= 3 \end{aligned}$$

It is a 3×4 system since it has 3 equations in 4 unknowns. Determine whether (*a*) $u = (-8, 6, 1, 1)$ and (*b*) $v = (-10, 5, 1, 2)$ are solutions of the system.

(*a*) Substitute the values of u in each equation, obtaining

$$\begin{aligned} -8+6+4(1)+3(1) &= -8+6+4+3=5 \\ 2(-8)+3(6)+1-2(1) &= -16+18+1-2=1 \\ -8+2(6)-5(1)+4(1) &= -8+12-5+4=3 \end{aligned}$$

Thus, u is a solution of the system since it is a solution of each equation.

(*b*) Substitute the values of v into each successive equation, obtaining

$$\begin{aligned} -10+5+4(1)+3(2) &= -10+5+4+6=5 \\ 2(-10)+3(5)+1-2(2) &= -20+15+1-4=-8 \neq 1 \\ -10+2(5)-5(1)+4(2) &= -10+10-5+8=3 \end{aligned}$$

Therefore, v is not a solution of the system, since it is not a solution of the second equation.

The system (3.2) of linear equations is said to be *consistent* if it has one or more solutions, and it is said to be *inconsistent* if it has no solution. If the field K of scalars is infinite, such as when K is the real field **R**, then we have the following important result.

Theorem 3.1: Suppose the field K is infinite. Then any system $\mathscr{L}$ of linear equations has either:
(i) a unique solution, (ii) no solution, or (iii) an infinite number of solutions.

This situation is pictured in Figure 3-1. The three cases have a geometrical description when the system $\mathscr{L}$ consists of two equations in two unknowns. These types of systems will be discussed later in this chapter.

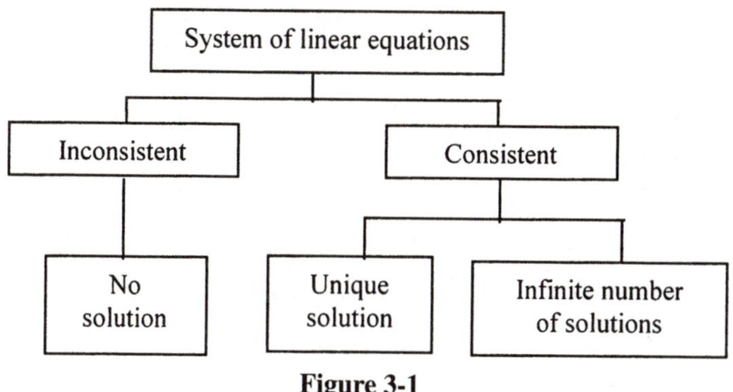

Figure 3-1

Augmented and Coefficient Matrices of a System

Consider again the general system (3.2) of m equations in n unknowns. Such a system has associated with it the following two matrices:

$$M = \begin{bmatrix} a_{11} & a_{12} & \cdots & a_{1n} & b_1 \\ a_{21} & a_{22} & \cdots & a_{2n} & b_2 \\ \cdots & \cdots & \cdots & \cdots & \cdots \\ a_{m1} & a_{m2} & \cdots & a_{mn} & b_n \end{bmatrix} \text{ and } A = \begin{bmatrix} a_{11} & a_{12} & \cdots & a_{1n} \\ a_{21} & a_{22} & \cdots & a_{2n} \\ \cdots & \cdots & \cdots & \cdots \\ a_{m1} & a_{m2} & \cdots & a_{mn} \end{bmatrix}$$

The first matrix M is called the *augmented matrix* of the system, and the second matrix A is called the *coefficient matrix.*

The coefficient matrix A is simply the matrix of coefficients, which is the augmented matrix M without the last column of constants. Some texts write $M = [A, B]$ to emphasize the two parts of M, where B denotes the column vector of constants. The augmented matrix M and the coefficient matrix A of the system in Example 3.2 are as follows:

$$M = \begin{bmatrix} 1 & 1 & 4 & 3 & 5 \\ 2 & 3 & 1 & -2 & 1 \\ 1 & 2 & -5 & 4 & 3 \end{bmatrix} \text{ and } A = \begin{bmatrix} 1 & 1 & 4 & 3 \\ 2 & 3 & 1 & -2 \\ 1 & 2 & -5 & 4 \end{bmatrix}$$

As expected, A consists of all the columns of M except the last, which is the column of constants.

Clearly, a system of linear equations is completely determined by its augmented matrix M, and vice versa. Specifically, each row of M corresponds to an equation of the system, and each column of M corresponds to the coefficients of an unknown, except for the last column, which corresponds to the constants of the system.

A linear equation is said to be *degenerate* if all the coefficients are zero, that is, if it has the form

$$0x_1 + 0x_2 + \ldots + 0x_n = b. \tag{3.3}$$

The solution of such an equation only depends on the value of the constant b. Specifically:

(i) If $b \neq 0$, then the equation has no solution.
(ii) If $b = 0$, then every vector $u = (k_1, k_2, \ldots, k_n)$ in K^n is a solution.

Theorem 3.2: Let $\mathscr{L}$ be a system of linear equations that contains a degenerate equation L, say with constant b.

(i) If $b \neq 0$, then the system $\mathscr{L}$ has no solution.
(ii) If $b = 0$, then L may be deleted from the system without changing the solution set of the system.

Part (i) comes from the fact that the degenerate equation has no solution, so the system has no solution. Part (ii) comes from the fact that every element in K^n is a solution of the degenerate equation.

Now let L be a nondegenerate linear equation. This means one or more of the coefficients of L are not zero. By the *leading unknown* of L, we mean the first unknown in L with a nonzero coefficient. For example, x_3 and y are the leading unknowns, respectively, in the equations $0x_1 + 0x_2 + 5x_3 + 6x_4 + 0x_5 + 8x_6 = 7$ and $0x + 2y - 4z = 5$.

We frequently omit terms with zero coefficients, so the above equations would be written as $5x_3 + 6x_4 + 8x_6 = 7$ and $2y - 4z = 5$. In such a case, the leading unknown appears first.

Equivalent Systems and Elementary Operations

Consider the system (3.2) of m linear equations in n unknowns. Let L be the linear equation obtained by multiplying the m equations by constants $c_1, c_2, .., c_m$, respectively and then adding the resulting equations. Specifically, let L be the following linear equation:

$$(c_1a_{11} + \ldots + c_ma_{m1})x_1 + \ldots + (c_1a_{1n} + \ldots + c_ma_{mn})x_n = c_1b_1 + \ldots + c_mb_m$$

Then L is called a *linear combination* of the equations in the system. One can easily show that any solution of the system (3.2) is also a solution of the linear combination L.

Example 3.3. Let L_1, L_2, L_3 denote, respectively, the three equations in Example 3.2. Let L be the equation obtained by multiplying L_1, L_2, L_3 by 3, −2, 4, respectively and then adding. Namely,

$$\begin{array}{rr} 3L_1: & 3x_1 + 3x_2 + 12x_3 + 9x_4 = 15 \\ -2L_2: & -4x_1 - 6x_2 - 2x_3 + 4x_4 = -2 \\ 4L_3: & 4x_1 + 8x_2 - 20x_3 + 16x_4 = 12 \\ \hline (\text{Sum})L: & 3x_1 + 5x_2 - 10x_3 + 29x_4 = 25 \end{array}$$

Then L is a linear combination of L_1, L_2, L_3. As expected, the solution $u = (-8, 6, 1,1)$ of the system is also a solution of L. That is substituting u in L, we obtain a true statement:

$$3(-8) + 5(6) - 10(1) + 29(1) = -24 + 30 - 10 + 29 = 25$$

Theorem 3.3: Two systems of linear equations have the same solutions if and only if each equation in each system is a linear combination of the equations in the other system.

Two systems of linear equations are said to be *equivalent* if they have the same solutions. The next subsection shows one way to obtain equivalent systems of linear equations.

Elementary Operations

The following operations on a system of linear equations $L_1, L_2, \ldots, L_m$ are called *elementary operations*.

[E_1] Interchange two of the equations. We indicate that the equations L_i and L_j are interchanged by writing:

$$\text{“Interchange } L_i \text{ and } L_j\text{”, or “}L_i \leftrightarrow L_j\text{”}$$

[E_2] Replace an equation by a nonzero multiple of itself. We indicate that equation L_i is replaced by kL_i (where $k \neq 0$) by writing:

$$\text{“Replace } L_i \text{ by } kL_i\text{”, or “}kL_i \rightarrow L_i\text{”}$$

[E_3] Replace an equation by the sum of a multiple of another equation and itself. We indicate that equation L_j is replaced by the sum of kL_i and L_j by writing:

$$\text{“Replace } L_j \text{ by } kL_i + L_j\text{” or “}kL_i + L_j \rightarrow L_j\text{”}$$

The arrow $\rightarrow$ in [E_2] and [E_3] may be read as “replaces.”

Sometimes we may apply [E_2] and [E_3] in one step:
[E] Replace equation L_j by the sum of kL_i and $k'L_j$, written

$$\text{“}kL_i + k'L_j \rightarrow L_j\text{”}$$

The main property of the above elementary operations is contained in the following theorem.

Theorem 3.4: Suppose a system $\mathcal{M}$ of linear equations is obtained from a system $\mathcal{L}$ of linear equations by a finite sequence of elementary operations. Then $\mathcal{M}$ and $\mathcal{L}$ have the same solutions.

Gaussian elimination, our main method for finding the solution of a given system of linear equations, consists of using the above operations to transform a given system into an equivalent system whose solution can be easily obtained. The details of Gaussian elimination are discussed in subsequent sections.

Small Square Systems of Linear Equations

This section considers the special case of one equation in one unknown, and two equations in two unknowns. These simple systems are treated separately since their solution sets can be described geometrically, and their properties motivate the general case.

Theorem 3.5: Consider the linear equation $ax = b$.
(i) If $a \neq 0$, then $x = b/a$ is a unique solution of $ax = b$.
(ii) If $a = 0$, but $b \neq 0$, then $ax = b$ has no solution.
(iii) If $a = 0$ and $b = 0$, then every scalar k is a solution of $ax = b$.

Example 3.4. Solve

$$(a)\ 4x - 1 = x + 6,\ (b)\ 2x - 5 - x = x + 3,\ (c)\ 4 + x - 3 = 2x + 1 - x.$$

(*a*) Rewrite the equation in standard form obtaining $3x = 7$. Then $x = \frac{7}{3}$ is the unique solution.
(*b*) Rewrite the equation in standard form obtaining $0x = 8$. The equation has no solution.
(*c*) Rewrite the equation in standard form obtaining $0x = 0$. Then every scalar k is a solution.

Consider a system of two nondegenerate linear equations in two unknowns x and y, which can be put in the standard form

$$\begin{aligned} A_1x + B_1y &= C_1 \\ A_2x + B_2y &= C_2 \end{aligned} \tag{3.4}$$

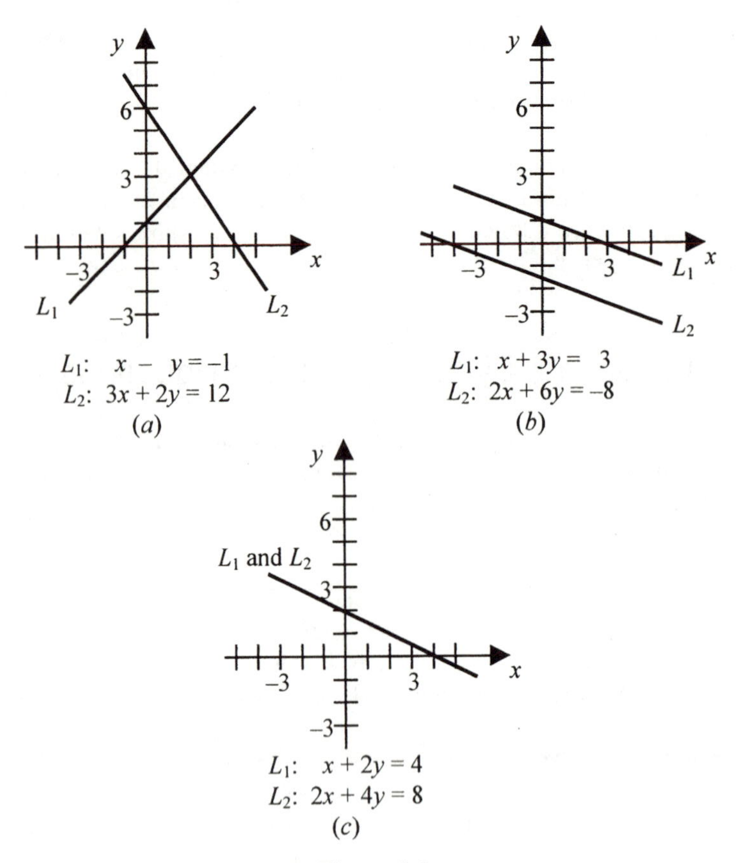

Figure 3-2

Since the equations are nondegenerate, A_1 and B_1 are not both zero, and A_2 and B_2 are not both zero.

The general solution of the system (3.4) belongs to one of three types as indicated in Figure 3-1. If **R** is the field of scalars, then the graph of each equation is a line in the plane $\mathbf{R}^2$ and the three types may be described geometrically as pictured in Figure 3-2. Specifically:

(1) *The system has exactly one solution.*
Here the two lines intersect in one point. This occurs when the lines have distinct slopes or, equivalently, when the coefficients of x and y are not proportional:

$$\frac{A_1}{A_2} \neq \frac{B_1}{B_2} \text{ or, equivalently, } A_1B_2 - A_2B_1 \neq 0$$

For example, in Figure 3-2 (a), $1/3 \neq -1/2$.

(2) *The system has no solution.*
Here the two lines are parallel. This occurs when the lines have the same slopes but different y intercepts, or when

$$\frac{A_1}{A_2} = \frac{B_1}{B_2} \neq \frac{C_1}{C_2}$$

For example, in Figure 3-2(b) $1/2 = 3/6 \neq -3/8$.

(3) *The system has an infinite number of solutions.*
Here the two lines coincide. This occurs when the lines have the same slopes and same y intercepts, or when the coefficients and constants are proportional,

$$\frac{A_1}{A_2} = \frac{B_1}{B_2} = \frac{C_1}{C_2}$$

For example, in Figure 3-2(c) $1/2 = 2/4 = 4/8$.

Elimination Algorithm

The solution to system (3.4) can be obtained by the process of elimination, whereby we reduce the system to a single equation in only one unknown. Assuming the system has a unique solution, this elimination algorithm has two parts.

Algorithm 3.1: The input consists of two nondegenerate linear equations L_1 and L_2 in two unknowns with a unique solution.

Part A. (Forward Elimination) Multiply each equation by a constant so that the resulting coefficients of one unknown are negatives of each other, and then add the two equations to obtain a new equation L that has only one unknown.

Part B. (Back-substitution) Solve for the unknown in the new equation L (which contains only one unknown), substitute this value of the unknown into one of the original equations, and then solve to obtain the value of the other unknown.

Part A of Algorithm 3.1 can be applied to any system even if the system does not have a unique solution. In such a case, the new equation L will be degenerate and Part B will not apply.

Example 3.5. Solve the system

$$\begin{aligned} L_1&: \quad 2x - 3y = -8 \\ L_2&: \quad 3x + 4y = 5 \end{aligned}$$

The unknown x is eliminated from the equations by forming the new equation $L = -3L_1 + 2L_2$. That is, we multiply L_1 by -3 and L_2 by 2 and add the resulting equations as follows:

$$\begin{array}{rr} -3L_1: & -6x + 9y = 24 \\ 2L_2: & 6x + 8y = 10 \\ \hline \text{Addition:} & 17y = 34 \end{array}$$

We now solve the new equation for y, obtaining $y = 2$. We substitute $y = 2$ into one of the original equations, say L_1, and solve for the other unknown x, obtaining

$$\begin{aligned} 2x - 3(2) &= -8 \\ 2x - 6 &= -8 \\ 2x &= -2 \\ x &= -1 \end{aligned}$$

Thus $x = -1$, $y = 2$, or the point $u = (-1, 2)$ is the unique solution of the system. The unique solution is expected since $2/3 \neq -3/4$.

Example 3.6. Solve the system

$$\begin{aligned} L_1&: \quad x - 3y = 4 \\ L_2&: \quad -2x + 6y = 5 \end{aligned}$$

We eliminated x from the equations by multiplying L_1 by 2 and adding it to L_2, that is, by forming the new equation $L = 2L_1 + L_2$. This yields the degenerate equation $0x + 0y = 13$ which has a nonzero constant $b = 13$. Thus this equation and the system has no solution. This is expected, since $1/(-2) = -3/6 \neq 4/5$. (Geometrically, the lines corresponding to the equations are parallel.)

Example 3.7. Solve the system

$$\begin{aligned} L_1: &\quad x - 3y = 4 \\ L_2: &\quad -2x + 6y = -8 \end{aligned}$$

We eliminated x from the equations by multiplying L_1 by 2 and adding it to L_2, that is, by forming the new equation $L = 2L_1 + L_2$. This yields the degenerate equation $0x + 0y = 0$ where the constant term is also zero. Thus the system has an infinite number of solutions, which correspond to the solutions of either equation. This is expected, since $1/(-2) = -3/6 = 4/(-8)$. (Geometrically, the lines corresponding to the equations coincide.)

To find the general solution, let $y = a$, and substitute into L_1 to obtain $x - 3\text{a} = 4$ or $x = 3a + 4$. Thus the general solution of the system is $u = (3a + 4, a)$ where a (called a *parameter*) is any scalar.

Systems in Triangular and Echelon Form

The main method for solving systems of linear equations, Gaussian elimination, is treated in subsequent sections. Here we consider two simple types of systems of linear equations: systems in triangular form and the more general systems in echelon form.

Consider the following system of linear equations, which is in triangular form:

$$\begin{aligned} 2x_1 + 3x_2 + 5x_3 - 2x_4 &= 9 \\ 5x_2 - x_3 + 3x_4 &= 1 \\ 7x_3 - x_4 &= 3 \\ 2x_4 &= 8 \end{aligned}$$

That is, the first unknown x_1 is the leading unknown in the first equation, the second unknown x_2 is the leading unknown in the second equation, and so on. Thus, in particular, the system is square and each leading unknown is *directly* to the right of the leading unknown in the preceding equation.

Such a triangular system always has a unique solution, which may be obtained by *back-substitution*. That is,

(1) First solve the last equation for the last unknown to get $x_4 = 4$.
(2) Then substitute this value $x_4 = 4$ in the next-to-last equation, and solve for the next-to-last unknown x_3 as follows:

$$7x_3 - 4 = 3 \quad \text{or} \quad 7x_3 = 7 \quad \text{or} \quad x_3 = 1$$

(3) Now substitute $x_3 = 1$ and $x_4 = 4$ in the second equation, and solve for the second unknown x_2 as follows:

$$5x_2 - 1 + 12 = 1 \quad \text{or} \quad 5x_2 + 11 = 1 \quad \text{or} \quad 5x_2 = -10 \quad \text{or} \quad x_2 = -2$$

(4) Finally, substitute $x_2 = -2, x_3 = 1, x_4 = 4$ in the first equation, and solve for the first unknown x_1 as follows:

$$2x_1 - 6 + 5 - 8 = 9 \quad \text{or} \quad 2x_1 - 9 = 9 \quad \text{or} \quad 2x_1 = 18 \quad \text{or} \quad x_1 = 9$$

Thus $x_1 = 9$, $x_2 = -2$, $x_3 = 1$, $x_4 = 4$, or equivalently, the vector $u = (9, -2, 1, 4)$ is the unique solution of the system.

Echelon Form, Pivot and Free Variables

The following system of linear equations is said to be in *echelon form*:

$$\begin{aligned} 2x_1 + 6x_2 - x_3 + 4x_4 - 2x_5 &= 7 \\ x_3 + 2x_4 + 2x_5 &= 5 \\ 3x_4 - 9x_5 &= 6 \end{aligned}$$

That is, no equation is degenerate and the leading unknown in each equation other than the first is to the right of the leading unknown in the preceding equation. The leading unknowns in the system, x_1, x_3, x_4, are called *pivot* variables and the other unknowns, x_2 and x_5, are called *free* variables.

Generally speaking, an *echelon system* or a *system in echelon form* has the following form:

$$
\begin{aligned}
a_{11}x_1 + a_{12}x_2 + a_{13}x_3 + a_{14}x_4 + \ldots + a_{1n}x_n &= b_1 \\
a_{2j_2}x_{j_2} + a_{2j_2+1}x_{j_2+1} + \ldots + a_{2n}x_n &= b_2 \\
\cdots\cdots\cdots\cdots\cdots\cdots\cdots\cdots\cdots & \\
a_{rj_r}x_{j_r} + \ldots + a_{rn}x_n &= b_r
\end{aligned}
$$

where $1 < j_2 < \ldots < j_r$ and all $a_{11}, a_{2j_2}, \ldots, a_{rj_r}$, are not zero. The *pivot* variables are $x_1, x_{j_2}, \ldots, x_{j_r}$. Note that $r \leq n$.

The solution set of any echelon system is described in the following theorem.

Theorem 3.6: Consider a system of linear equations in echelon form, say with r equations in n unknowns. There are two cases.

(i) $r = n$. That is, there are as many equations as unknowns (triangular form). Then the system has a unique solution.

(ii) $r < n$. That is, there are more unknowns than equations. Then we can arbitrarily assign values to the $n - r$ free variables and solve uniquely for the r pivot variables, obtaining a solution of the system.

Gaussian Elimination

The main method for solving the general system (3.2) of linear equations is called *Gaussian elimination*. It essentially consists of two parts:

Part A. (Forward Elimination) Step-by-step reduction of the system yielding either a degenerate equation with no solution (which indicates the system has no solution) or an equivalent simpler system in triangular or echelon form.

Part B. (Backward Elimination) Step-by-step back-substitution to find the solution of the simpler system.

Part B has already been investigated in previous sections. Accordingly, we need only give the algorithm for Part A, which is as follows.

Algorithm 3.2 (for Part A):

Input: The $m \times n$ system (3.2) of linear equations.

Elimination Step: Find the first unknown in the system with a nonzero coefficient (which now must be x_1).

(*a*) Arrange so that $a_{11} \neq 0$. That is, if necessary, interchange equations so that the first unknown x_1 appears with a nonzero coefficient in the first equation.

(*b*) Use a_{11} as a pivot to eliminate x_1 from all equations except the first equation. That is, for $i > 1$:

(1) Set $m = -a_{i1}/a_{11}$; (2) Replace L_i by $mL_1 + L_i$ or replace L_i by $-a_{i1}L_1 + a_{11}L_i$

The system now has the following form:

$$\begin{aligned} a_{11}x_1 + a_{12}x_2 + a_{13}x_3 + a_{14}x_4 + \ldots + a_{1n}x_n &= b_1 \\ a_{2j_2}x_{j_2} + a_{2j_2+1}x_{j_2+1} + \ldots + a_{2n}x_n &= b_2 \\ \ldots\ldots\ldots\ldots\ldots\ldots\ldots\ldots\ldots\ldots\ldots \\ a_{mj_2}x_{j_2} + \ldots + a_{mn}x_n &= b_n \end{aligned}$$

where x_1 does not appear in any equation except the first, $a_{11} \neq 0$, and denotes the first unknown with a nonzero coefficient in any equation other than the first. (The number m is known as the *multiplier*.)

(*c*) Examine each new equation L.

(1) If L has the form $0x_1 + 0x_2 + \ldots + 0x_n = b$ with $b \neq 0$, then

STOP

The system is *inconsistent* and has no solution.

(2) If L has the form $0x_1 + 0x_2 + \ldots + 0x_n = 0$ or if L is a multiple of another equation, then delete L from the system.

Recursion Step: Repeat the Elimination Step with each new "smaller" subsystem formed by all the equations excluding the first equation.

Output: Finally, the system is reduced to triangular or echelon form, or a degenerate equation with no solution is obtained indicating an inconsistent system.

Example 3.8. Solve the following system by Gaussian elimination.

$$\begin{aligned} L_1: &\quad x - 3y - 2z = 6 \\ L_2: &\quad 2x - 4y - 3z = 8 \\ L_3: &\quad -3x + 6y + 8z = -5 \end{aligned}$$

Part A. We use the coefficient 1 of x in the first equation L_1 as the pivot in order to eliminate x from the second equation L_2 and from the third equation L_3. This is accomplished as follows:

(1) Multiply L_1 by the multiplier $m = -2$ and add it to L_2; that is, "Replace L_2 by $-2L_1 + L_2$."
(2) Multiply L_1 by the multiplier $m = 3$ and add it to L_3; that is, "Replace L_3 by $3L_1 + L_3$."
These steps yield

$$\begin{array}{lrr} (-2)\,L_1: & -2x + 6y + 4z = & -12 \\ L_2: & 2x - 4y - 3z = & 8 \\ \hline \text{New } L_2: & 2y + z = & -4 \end{array} \qquad \begin{array}{lrr} 3\,L_1: & 3x - 9y - 6z = & 18 \\ L_3: & -3x + 6y + 8z = & -5 \\ \hline \text{New } L_3: & -3y + 2z = & 13 \end{array}$$

Thus the original system is replaced by the following system:

$$\begin{aligned} L_1: &\quad x - 3y - 2z = 6 \\ L_2: &\quad 2y + z = -4 \\ L_3: &\quad -3y + 2z = 13 \end{aligned}$$

Next we use the coefficient 2 of y in the (new) second equation L_2 as the pivot in order to eliminate y from the (new) third equation L_3. This is accomplished as follows:

(3) Multiply L_2 by the multiplier $m = \frac{3}{2}$ and add it to L_3; that is, "Replace L_3 by $\frac{3}{2} L_2 + L_3$" (alternately, "Replace L_3 by $3L_2 + 2L_3$," which will avoid fractions).
This step yields

$\frac{3}{2}L_2$:	$3y + \frac{3}{2}z = -6$		$3L_2$:	$6y + 3z = 12$	
L_3:	$-3y + 2z = 13$	or	$2L_3$:	$-6y + 4z = 26$	
New L_3:	$7/2z = 7$		New L_3:	$7z = 14$	

Thus our system is replaced by the following system:

$$\begin{aligned} L_1&: & x - 3y - 2z &= 6 \\ L_2&: & 2y + z &= -4 \\ L_3&: & 7z &= 14 \quad (\text{or } 7/2z = 7) \end{aligned}$$

The system is now in triangular form, so Part A is completed.

Part B. The values for the unknowns are obtained in reverse order, z, y, x by back-substitution. Specifically:

(1) Solve for z in L_3 to get $z = 2$.
(2) Substitute $z = 2$ in L_2, and solve for y to get $y = -3$.
(3) Substitute $z = 2$ and $y = -3$ in L_1 and solve for x to get $x = 1$.

Thus the solution of the system is $x = 1$, $y = -3$, $z = 2$ or, equivalently, $u = (1, -3, 2)$

The Gaussian elimination algorithm involves rewriting systems of linear equations. Sometimes we can avoid excessive recopying of some of the equations by adopting a "condensed format." This format for the solution of the above system follows:

Number	Equation	Operation
(1)	$x - 3y - 2z = 6$	
(2)	$2x - 4y - 3z = 8$	
(3)	$-3x + 6y + 8z = -5$	
(2′)	$2y + z = -4$	Replace L_2 by $-2L_1 + L_2$
(3′)	$-3y + 2z = 13$	Replace L_3 by $3L_1 + L_3$
(3″)	$7z = 14$	Replace L_3 by $3L_2 + 2L_3$

That is, first we write down the number of each of the original equations. As we apply the Gaussian elimination algorithm to the system, we only write down the new equations, and we label each new equation using the same number as the original corresponding equation, but with an added prime. (After each new equation, we will indicate, for instructional purposes, the elementary operation that yielded the new equation.)

The system in triangular form consists of equations (1), (2′), and (3″), the numbers with the largest number of primes. Applying back-substitution to these equations again yields $x = 1$, $y = -3$, $z = 2$.

Remember

If two equations need to be interchanged, say to obtain a nonzero coefficient as a pivot, then simply renumber the two equations rather than change their positions.

Example 3.9. Solve the following system:

$$\begin{aligned} x + 2y - 3z &= 1 \\ 2x + 5y - 8z &= 4 \\ 3x + 8y - 13z &= 7 \end{aligned}$$

We solve by Gaussian elimination.

Part A. (Forward Elimination)

Number	Equation	Operation
(1)	$x + 2y - 3z = 1$	
(2)	$2x + 5y - 8z = 4$	
(3)	$3x + 8y - 13z = 7$	
(2′)	$y - 2z = 2$	Replace L_2 by $-2L_1 + L_2$
(3′)	$2y - 4z = 4$	Replace L_3 by $-3L_1 + L_3$
(3″)	$0 = 0$	Replace L_3 by $-2L_2 + L_3$

Notice that the third equation is deleted since it is a multiple of the second equation.

$$\begin{aligned} x + 2y - 3z &= 1 \\ y - 2z &= 2 \end{aligned}$$

The system is now in echelon form with free variable z:

Part B. (Backward Elimination) To obtain the general solution, let the free variable $z = a$, and solve for x and y by back-substitution. Substitute $z = a$ in the second equation to obtain $y = 2 + 2a$. Then substitute $z = a$ and $y = 2 + 2a$ in the first equation to obtain $x = -3 - a$.

Thus the following is the general solution where a is a parameter:

$$x = -3 - a,\ y = 2 + 2a,\ z = \text{a} \text{ or } u = (-3 - a, 2 + 2a, a).$$

Example 3.10. Solve the following system:

$$\begin{aligned} x_1 + 3x_2 - 2x_3 + 5x_4 &= 4 \\ 2x_1 + 8x_2 - x_3 + 9x_4 &= 9 \\ 3x_1 + 5x_2 - 12x_3 + 17x_4 &= 7 \end{aligned}$$

Number	Equation	Operation
(1)	$x_1 + 3x_2 - 2x_3 + 5x_4 = 4$	
(2)	$2x_1 + 8x_2 - x_3 + 9x_4 = 9$	
(3)	$3x_1 + 5x_2 - 12x_3 + 17x_4 = 7$	
(2′)	$2x_2 + 3x_3 - x_4 = 1$	$-2L_1 + L_2 \rightarrow L_2$
(3′)	$-4x_2 - 6x_3 + 2x_4 = -5$	$-3L_1 + L_3 \rightarrow L_3$
(3″)	$0 = -3$	$2L_2 + L_3 \rightarrow L_3$

This last equation, and hence the original system have no solution.

Echelon Matrices; Row Canonical Form

One way to solve a system of linear equations is by working with its augmented matrix M rather than the system itself. This section introduces the necessary matrix concepts for such a discussion. These concepts, such as echelon matrices and elementary row operations, are also of independent interest.

A matrix A is called an *echelon matrix*, or is said to be in *echelon form*, if the following two conditions hold (where a *leading nonzero element* of a row of A is the first nonzero element in the row):

(1) All zero rows, if any, are at the bottom of the matrix.
(2) Each leading nonzero entry in a row is to the right of the leading nonzero entry in the preceding row.

That is, $A = [a_{ij}]$ is an echelon matrix if there exist nonzero entries

$$a_{1j_1}, a_{2j_2}, \ldots, a_{rj_r}, \quad \text{where } j_1 < j_2 < \ldots < j_r$$

with the property that

$$a_{ij} = 0 \text{ for } \begin{cases} i \le r, & j < j_r \\ i > r \end{cases}$$

The entries $a_{1j_1}, a_{2j_2}, \ldots, a_{rj_r}$, which are the leading nonzero elements in their respective rows, are called the *pivots* of the echelon matrix.

Example 3.11. The following is an echelon matrix whose pivots have been shaded:

$$A = \begin{bmatrix} 0 & \boxed{2} & 3 & 4 & 5 & 9 & 0 & 7 \\ 0 & 0 & 0 & \boxed{3} & 4 & 1 & 2 & 5 \\ 0 & 0 & 0 & 0 & 0 & \boxed{5} & 7 & 2 \\ 0 & 0 & 0 & 0 & 0 & 0 & \boxed{8} & 6 \\ 0 & 0 & 0 & 0 & 0 & 0 & 0 & 0 \end{bmatrix}$$

Observe that the pivots are in columns C_2, C_4, C_6, C_7, and each is to the right of the one above. Using the above notation, the pivots are

$$a_{1j_1} = 2, \quad a_{2j_2} = 3, \quad a_{3j_3} = 5, \quad a_{4j_4} = 8$$

where $j_1 = 2, j_2 = 4, j_3 = 6, j_4 = 7$. Here $r = 4$.

A matrix A is said to be in *row canonical form* if it is an echelon matrix, that is, if it satisfies the above properties (1) and (2), and if it satisfies the following additional two properties:

(3) Each pivot (leading nonzero entry) is equal to 1.
(4) Each pivot is the only nonzero entry in its column.

The major difference between an echelon matrix and a matrix in row canonical form is that in an echelon matrix there must be zeros below the pivots [Properties (1) and (2)], but in a matrix in canonical form, each pivot must also equal 1[Property 3] and there must also be zeros above the pivots [Property (4)].

The zero matrix 0 of any size and the identity matrix I of any size are important special examples of matrices in row canonical form.

Example 3.12. The following are echelon matrices whose pivots have been shaded:

$$\begin{bmatrix} 2 & 3 & 2 & 0 & 4 & 5 & -6 \\ 0 & 0 & 1 & 1 & -3 & 2 & 0 \\ 0 & 0 & 0 & 0 & 0 & 6 & 2 \\ 0 & 0 & 0 & 0 & 0 & 0 & 0 \end{bmatrix}, \quad \begin{bmatrix} 1 & 3 & 2 \\ 0 & 0 & 1 \\ 0 & 0 & 0 \end{bmatrix}, \quad \begin{bmatrix} 0 & 1 & 3 & 0 & 0 & 4 \\ 0 & 0 & 0 & 1 & 0 & -3 \\ 0 & 0 & 0 & 0 & 1 & 2 \end{bmatrix}$$

The third matrix is also an example of a matrix in row canonical form. The second matrix is not in row canonical form, since it does not satisfy property (4), that is, there is a nonzero entry above the second pivot in the third column. The first matrix is not in row canonical form, since it satisfies neither property (3) nor property (4), that is, some pivots are not equal to 1 and there are nonzero entries above the pivots.

Suppose A is a matrix with rows $R_1, R_2, \ldots, R_m$. The following operations on A are called *elementary row operations*.

[E_1] Interchange rows R_i and R_j. This may be written as:

"Interchange R_i and R_j," or "$R_i \leftrightarrow R_j$"

[E_2] Replace row R_i by a nonzero multiple kR_i of itself. This may be written as:

"Replace R_i by kR_i $(k \neq 0)$," or "$kR_i \rightarrow R_i$"

[E_3] Replace row R_j by the sum of a multiple kR_i of a row R_i and itself. This may be written as:

$$\text{"Replace } R_j \text{ by } kR_i + R_j\text{," or "} kR_i + R_j \to R_j\text{"}$$

The arrow → in [E_2] and [E_3] may be read as "replaces."

Sometimes (say to avoid fractions when all the given scalars are integers) we may apply [E_2] and [E_3] in one step, that is, we may apply the following operation:

[E] Replace equation R_j by the sum of kR_i and $k'R_j$, written

$$\text{"}kR_i + k'R_j \to R_j\text{"}$$

We emphasize that in operations [E_3] and [E] only row R_j is changed.

A matrix A is said to be *row equivalent* to a matrix B, written

$$\mathrm{A} \sim \mathrm{B}$$

if B can be obtained from A by a sequence of elementary row operations. In the case that B is also an echelon matrix, B is called an *echelon form* of A.

Theorem 3.7: Suppose $A = [a_{ij}]$ and $B = [b_{ij}]$ are row equivalent echelon matrices with respective pivot entries

$$a_{1j_1}, a_{2j_2}, \ldots, a_{rj_r} \text{ and } b_{1k_1}, b_{2k_2}, \ldots, b_{sk_s}$$

Then A and B have the same number of nonzero rows, that is, $r = s$, and the pivot entries are in the same positions, that is

$$j_1 = k_1, j_2 = k_2, \ldots, j_r = k_r.$$

Theorem 3.8: Every matrix A is row equivalent to a unique matrix in row canonical form. This unique matrix is called the row canonical form of A.

The *rank* of a matrix A, written rank(A), is equal to the number of pivots in an echelon form of A.

Gaussian Elimination (Matrices)

In this section we will introduce two new algorithms. These algorithms, which use the elementary row operations, are simply restatements of *Gaussian elimination* as applied to matrices rather than to linear equations. (The term "row reduce" or simply "reduce" will mean to transform a matrix by the elementary row operations.)

Algorithm 3.3 (**Forward Elimination**): The input is any matrix A. (The algorithm puts 0's below each pivot, working from the "top-down.") The output is an echelon form of A.

Step 1: Find the first column with a nonzero entry. Let j_1 denote this column.

(*a*) Arrange so that $a_{1j_1} \neq 0$. That is, if necessary, interchange rows so that a nonzero entry appears in the first row in column j_1.

(*b*) Use a_{1j_1} as a pivot to obtain 0's below a_{1j_1}. Specifically, for $i > 1$:

(1) Set $m = -a_{ij_1} / a_{1j_1}$; (2) Replace R_i by $mR_1 + R_i$ or replace R_i by $-a_{ij_1} R_1 + a_{1j_1} R_i$

(The number m is known as the *multiplier*.)

Step 2: Repeat Step 1 with the submatrix formed by all the rows excluding the first row. Here we let j_2 denote the first column in the subsystem with a nonzero entry. Hence, at the end of Step 2, we have $a_{2j_2} \neq 0$.

Steps 3 to *r*. Continue the above process until a submatrix has only zero rows.

We emphasize that, at the end of the algorithm, the pivots will be $a_{1j_1}, a_{2j2}, \ldots, a_{rj_r}$ where r denotes the number of nonzero rows in the final echelon matrix.

Algorithm 3.4 (Backward Elimination): The input is a matrix $A = [a_{ij}]$ in echelon form with pivot entries $a_{1j_1}, a_{2j_2}, \ldots, a_{rj_r}$ the output is the row canonical form of A.

Step 1. (*a*) (Use row scaling so the last pivot equals 1.) Multiply the last nonzero row R_r by $1/a_{rj_r}$.

(*b*) (Use $a_{rj_r} = 1$ to obtain 0's above the pivot.) For $i = r-1, r-2, \ldots, 2, 1$;

(1) Set $m = a_{rj_r}$; (2) Replace R_i by $mR_r + R_i$

Step 2 to $r - 1$. Repeat Step 1 for rows $R_{r-1}, R_{r-2}, \ldots, R_2$.

Step r. (Use row scaling so the first pivot equals 1.) Multiply R_1 by $1/a_{1j_1}$.

> Remember that Gaussian elimination is a two-stage process! **Stage A** (Algorithm 3.3) puts 0's below each pivot and **Stage B** (Algorithm 3.4) puts 0's above each pivot.

Example 3.13. Consider the matrix $A = \begin{bmatrix} 1 & 2 & -3 & 1 & 2 \\ 2 & 4 & -4 & 6 & 10 \\ 3 & 6 & -6 & 9 & 13 \end{bmatrix}$.

(*a*) Use Algorithm 3.3 to reduce A to an echelon form.

(*b*) Use Algorithm 3.4 to further reduce A to its row canonical form.

(*a*) First use $a_{11} = 1$ as a pivot to obtain 0's below a_{11}, that is, apply the operations "Replace R_2 by $-2R_1 + R_2$" and "Replace R_3 by $-3R_1 + R_3$"; and then use $a_{23} = 2$ as a pivot to obtain 0 below a_{23}, that is, apply the operation "Replace R_3 by $-\frac{3}{2}R_2 + R_3$." This yields

$$A \sim \begin{bmatrix} 1 & 2 & -3 & 1 & 2 \\ 0 & 0 & 2 & 4 & 6 \\ 0 & 0 & 3 & 6 & 7 \end{bmatrix} \sim \begin{bmatrix} 1 & 2 & -3 & 1 & 2 \\ 0 & 0 & 2 & 4 & 6 \\ 0 & 0 & 0 & 0 & -2 \end{bmatrix}$$

The matrix is now in echelon form.

(*b*) Multiply R_3 by $-\frac{1}{2}$ so the pivot entry $a_{35} = 1$, and then use $a_{35} = 1$ as a pivot to obtain 0's above it by the operations "Replace R_2 by $-5R_3 + R_2$" and then "Replace R_1 by $-2R_3 + R_1$." This yields

$$A \sim \begin{bmatrix} 1 & 2 & -3 & 1 & 2 \\ 0 & 0 & 2 & 4 & 6 \\ 0 & 0 & 0 & 0 & 1 \end{bmatrix} \sim \begin{bmatrix} 1 & 2 & -3 & 1 & 0 \\ 0 & 0 & 2 & 4 & 0 \\ 0 & 0 & 0 & 0 & 1 \end{bmatrix}$$

Multiply R_2 by $\frac{1}{2}$ so the pivot entry $a_{23} = 1$, and then use $a_{23} = 1$ as a pivot to obtain 0's above it by the operation "Replace R_1 by $3R_2 + R_1$." This yields

$$A \sim \begin{bmatrix} 1 & 2 & -3 & 1 & 0 \\ 0 & 0 & 1 & 2 & 0 \\ 0 & 0 & 0 & 0 & 1 \end{bmatrix} \sim \begin{bmatrix} 1 & 2 & 0 & 7 & 0 \\ 0 & 0 & 1 & 2 & 0 \\ 0 & 0 & 0 & 0 & 1 \end{bmatrix}$$

The last matrix is the row canonical form of A.

Application to Systems of Linear Equations

One way to solve a system of linear equations is by working with its augmented matrix M rather than the equations themselves. Specifically, we reduce M to echelon form (which tells us whether the system has a solution), and then further reduce M to its row canonical form (which essentially gives the solution of the original system of linear equations).

Example 3.14. Solve each of the following systems:

$$\begin{array}{lll} x_1 + x_2 - 2x_3 + 4x_4 = 5 & x_1 + x_2 - 2x_3 + 3x_4 = 4 & x + 2y + z = 3 \\ 2x_1 + 2x_2 - 3x_3 + x_4 = 3 & 2x_1 + 3x_2 + 3x_3 - x_4 = 3 & 2x + 5y - z = -4 \\ 3x_1 + 3x_2 - 4x_3 - 2x_4 = 1 & 5x_1 + 7x_2 + 4x_3 + x_4 = 5 & 3x - 2y - z = 5 \\ (a) & (b) & (c) \end{array}$$

(*a*) Reduce its augmented matrix M to echelon form and then to row canonical form as follows:

$$M = \begin{bmatrix} 1 & 1 & -2 & 4 & 5 \\ 2 & 2 & -3 & 1 & 3 \\ 3 & 3 & -4 & -2 & 1 \end{bmatrix} \sim \begin{bmatrix} 1 & 1 & -2 & 4 & 5 \\ 0 & 0 & 1 & -7 & -7 \\ 0 & 0 & 2 & -14 & -14 \end{bmatrix}$$

$$\sim \begin{bmatrix} 1 & 1 & 0 & -10 & -9 \\ 0 & 0 & 1 & -7 & -7 \\ 0 & 0 & 0 & 0 & 0 \end{bmatrix}$$

Rewrite the row canonical form in terms of a system of linear equations to obtain the free variable form of the solution. That is,

$$\begin{aligned} x_1 + x_2 - 10x_4 &= -9 \\ x_3 - 7x_4 &= -7 \end{aligned} \quad \text{or} \quad \begin{aligned} x_1 &= -9 - x_2 + 10x_4 \\ x_3 &= -7 + 7x_4 \end{aligned}$$

(The zero row is omitted in the solution.) Observe that x_1 and x_3 are the pivot variables, and x_2 and x_4 are the free variables.

(*b*) First reduce its augmented matrix M to echelon form as follows:

$$M = \begin{bmatrix} 1 & 1 & -2 & 3 & 4 \\ 2 & 3 & 3 & -1 & 3 \\ 5 & 7 & 4 & 1 & 5 \end{bmatrix} \sim \begin{bmatrix} 1 & 1 & -2 & 3 & 4 \\ 0 & 1 & 7 & -7 & -5 \\ 0 & 2 & 14 & -14 & -15 \end{bmatrix}$$

$$\sim \begin{bmatrix} 1 & 1 & -2 & 3 & 4 \\ 0 & 1 & 7 & -7 & -5 \\ 0 & 0 & 0 & 0 & -5 \end{bmatrix}$$

There is no need to continue to find the row canonical form of M, since the echelon form already tells us that the system has no solution. Specifically, the third row of the echelon matrix corresponds to the degenerate equation $0x_1 + 0x_2 + 0x_3 + 0x_4 = -5$ which has no solution. Thus the system has no solution.

(*c*) Reduce its augmented matrix M to echelon form and then to row canonical form as follows:

$$M = \begin{bmatrix} 1 & 2 & 1 & 3 \\ 2 & 5 & -1 & -4 \\ 3 & -2 & -1 & 5 \end{bmatrix} \sim \begin{bmatrix} 1 & 2 & 1 & 3 \\ 0 & 1 & -3 & -10 \\ 0 & -8 & -4 & -4 \end{bmatrix}$$

$$\sim \begin{bmatrix} 1 & 2 & 1 & 3 \\ 0 & 1 & -3 & -10 \\ 0 & 0 & -28 & -84 \end{bmatrix} \sim \begin{bmatrix} 1 & 2 & 1 & 3 \\ 0 & 1 & -3 & -10 \\ 0 & 0 & 1 & 3 \end{bmatrix}$$

$$\sim \begin{bmatrix} 1 & 2 & 0 & 0 \\ 0 & 1 & 0 & -1 \\ 0 & 0 & 1 & 3 \end{bmatrix} \sim \begin{bmatrix} 1 & 0 & 0 & 2 \\ 0 & 1 & 0 & -1 \\ 0 & 0 & 1 & 3 \end{bmatrix}$$

Thus the system has the unique solution $x = 2, y = -1, z = 3$, or equivalently, the vector $u = (2, -1, 3)$.

The general system (3.2) of m linear equations in n unknowns is equivalent to the matrix equation

$$\begin{bmatrix} a_{11} & a_{12} & \cdots & a_{1n} \\ a_{21} & a_{22} & \cdots & a_{2n} \\ \cdots & \cdots & \cdots & \cdots \\ a_{m1} & a_{m2} & \cdots & a_{mn} \end{bmatrix} \begin{bmatrix} x_1 \\ x_2 \\ \cdots \\ x_n \end{bmatrix} = \begin{bmatrix} b_1 \\ b_2 \\ \cdots \\ b_m \end{bmatrix} \quad \text{or } AX = B$$

where $A = [a_{ij}]$ is the coefficient matrix, $X = [x_j]$ is the column vector of unknowns, and $B = [b_i]$ is the column vector of constants. The statement that the system of linear equations and the matrix equation are equivalent means that any vector solution of the system is a solution of the matrix equation and vice versa.

Example 3.15. The following system of linear equations and matrix equation are equivalent:

$$\begin{aligned} x_1 + 2x_2 - 4x_3 + 7x_4 &= 4 \\ 3x_1 - 5x_2 + 6x_3 - 8x_4 &= 8 \\ 4x_1 - 3x_2 - 2x_3 + 6x_4 &= 11 \end{aligned} \quad \text{and} \quad \begin{bmatrix} 1 & 2 & -4 & 7 \\ 3 & -5 & 6 & -8 \\ 4 & -3 & -2 & 6 \end{bmatrix} \begin{bmatrix} x_1 \\ x_2 \\ x_3 \\ x_4 \end{bmatrix} = \begin{bmatrix} 4 \\ 8 \\ 11 \end{bmatrix}$$

We note that $x_1 = 3, x_2 = 1, x_3 = 2, x_4 = 1$, or, in other words, the vector $u = [3, 1, 2, 1]$ is a solution of the system. Thus the (column) vector u is also a solution of the matrix equation.

The matrix form $AX = B$ of a system of linear equations is notationally very convenient when discussing and proving properties of systems of linear equations.

A system $AX = B$ of linear equations is square if and only if the matrix A of coefficients is square. In such a case, we have the following important result.

Theorem 3.9: A square system $AX = B$ of linear equations has a unique solution if and only if the matrix A is invertible. In such a case, $A^{-1}B$ is the unique solution of the system.

Example 3.16. Consider the following system of linear equations, whose coefficient matrix A and inverse A^{-1} are also given:

$$\begin{aligned} x + 2y + 3z &= 1 \\ x + 3y + 6z &= 3, \\ 2x + 6y + 13z &= 5 \end{aligned} \quad A = \begin{bmatrix} 1 & 2 & 3 \\ 1 & 3 & 6 \\ 2 & 6 & 13 \end{bmatrix}, \quad A^{-1} = \begin{bmatrix} 3 & -8 & 3 \\ -1 & 7 & -3 \\ 0 & -2 & 1 \end{bmatrix}$$

By Theorem 3.9, the unique solution of the system is

$$A^{-1}B = \begin{bmatrix} 3 & -8 & 3 \\ -1 & 7 & -3 \\ 0 & -2 & 1 \end{bmatrix} \begin{bmatrix} 1 \\ 3 \\ 5 \end{bmatrix} = \begin{bmatrix} -6 \\ 5 \\ -1 \end{bmatrix}$$

That is, $x = -6, y = 5, z = -1$.

We emphasize that Theorem 3.9 does not usually help us to find the solution of a square system. That is, finding the inverse of a coefficient matrix A is not usually any easier than solving the system directly. Thus, unless we are given the inverse of a coefficient matrix A, as in Example 3.16, we usually solve a square system by Gaussian elimination.

Inverse Matrices

The following algorithm finds the inverse of a matrix.

Algorithm 3.5: The input is a square matrix A. The output is the inverse of A or that the inverse does not exist.

Step 1. Form the $n \times 2n$ (block) matrix $M = [A, I]$, where A is the left half of M and the identity matrix I is the right half of M.

Step 2. Row reduce M to echelon form. If the process generates a zero row in the A half of M, then A has no inverse. (Otherwise A is in triangular form.)

Step 3. Further row reduce M to its row canonical form $M \sim [I, B]$ where the identity matrix I has replaced A in the left half of M.

Step 4. Set $A^{-1} = B$, the matrix that is now in the right half of M.

Example 3.16. Find the inverse of the matrix $A = \begin{bmatrix} 1 & 0 & 2 \\ 2 & -1 & 3 \\ 4 & 1 & 8 \end{bmatrix}$.

First form the (block) matrix $M = [A, I]$ and row reduce M to an echelon form:

$$M = \left[\begin{array}{ccc|ccc} 1 & 0 & 2 & 1 & 0 & 0 \\ 2 & -1 & 3 & 0 & 1 & 0 \\ 4 & 1 & 8 & 0 & 0 & 1 \end{array}\right] \sim \left[\begin{array}{ccc|ccc} 1 & 0 & 2 & 1 & 0 & 0 \\ 0 & -1 & -1 & -2 & 1 & 0 \\ 0 & 1 & 0 & -4 & 0 & 1 \end{array}\right]$$

$$\sim \left[\begin{array}{ccc|ccc} 1 & 0 & 2 & 1 & 0 & 0 \\ 0 & -1 & -1 & -2 & 1 & 0 \\ 0 & 0 & -1 & -6 & 1 & 1 \end{array}\right] \sim \left[\begin{array}{ccc|ccc} 1 & 0 & 0 & -11 & 2 & 2 \\ 0 & 1 & 0 & -4 & 0 & 1 \\ 0 & 0 & 1 & 6 & -1 & -1 \end{array}\right]$$

The identity matrix is now in the left half of the final matrix; hence the right half is A^{-1}. In other words,

$$A^{-1} = \begin{bmatrix} -11 & 2 & 2 \\ -4 & 0 & 1 \\ 6 & -1 & -1 \end{bmatrix}$$

Chapter 4
Vector Spaces

In This Chapter:

- ✔ *Notation*
- ✔ *Vector Spaces*
- ✔ *Examples of Vector Spaces*
- ✔ *Linear Combinations; Spanning Sets*
- ✔ *Subspaces*
- ✔ *Linear Span; Row Space of a Matrix*
- ✔ *Linear Dependence and Independence*
- ✔ *Basis and Dimension*
- ✔ *Rank of a Matrix*
- ✔ *Sums and Direct Sums*

Notation

This chapter introduces the underlying structure of linear algebra, that of a finite-dimensional vector space. The definition of a vector space V, whose elements are called *vectors*, involves an arbitrary field K, whose elements are called *scalars*. The following notation will be used (unless otherwise stated or implied):

V	the given vector space
u, v, w	vectors in V
K	the given number field
a, b, c or k	scalars in K

Almost nothing essential is lost if the reader assumes that K is the real field **R**. The reader might suspect that the real line **R** has "dimension" one, the Cartesian plane $\mathbf{R}^2$ has "dimension" two, and the space $\mathbf{R}^3$ has "dimension" three. This chapter formalizes the notation of "dimension," and this definition will agree with the reader's intuition.

Throughout this text, we will use the following set notation:

$a \in A$	Element a belongs to set A
$a, b \in A$	Element a and b belongs to set A
$\forall x \in A$	For every x in A
$\exists x \in A$	There exists an x in A
$A \subseteq B$	A is a subset of B
$A \cap B$	Intersection of A and B
$A \cup B$	Union of A and B
$\varnothing$	Empty set

Vector Spaces

Let V be a nonempty set with two operations:

(i) **Vector Addition**: This assigns to any $u, v \in V$ a *sum* $u + v$ in V.
(ii) **Scalar Multiplication**: This assigns to any $u \in V$, $k \in K$ a *product* $ku \in V$.

Then V is called a *vector space* (over the field K) if the following axioms hold for any vectors $u, v, w \in V$:

[A_1] $(u + v) + w = u + (v + w)$
[A_2] There is a vector in V, denoted by 0 and called the *zero vector*, such that, for any $u \in V$, $u + 0 = 0 + u = 0$

[A_3] For each $u \in V$, there is a vector in V, denoted by $-u$, and called the *negative* of u, such that $u + (-u) = (-u) + u = 0$.

[A_4] $u + v = v + u$.

[M_1] $k(u + v) = ku + kv$, for any scalar $k \in K$.

[M_2] $(a + b)u = au + bu$, for any scalars $a, b \in K$

[M_3] $(ab)u = a(bu)$, for any scalars $a, b \in K$.

[M_4] $1u = u$, for the unit scalar $1 \in K$.

The above axioms naturally split into two sets (as indicated by the labeling of the axioms). The first four are only concerned with the additive structure of V, and can be summarized by saying V is a *commutative group* under addition. This means:

(*a*) Any sum $v_1 + v_2 + \ldots + v_m$ of vectors requires no parentheses and does not depend on the order of the summands.

(*b*) The zero vector 0 is unique, and the negative $-u$ of a vector u is unique.

(*c*) (Cancellation Law) If $u + w = v + w$, then $u = v$.

Also, *subtraction* in V is defined by $u - v = u + (-v)$, where $-v$ is the unique negative of v.

Theorem 4.1: Let V be a vector space over a field K.

(i) For any scalar $k \in K$ and $0 \in V$, $k0 = 0$.

(ii) For $0 \in K$ and any vector $u \in V$, $0u = 0$.

(iii) If $ku = 0$, where $k \in K$ and $u \in V$, then $k = 0$ or $u = 0$.

(iv) For any $k \in K$ and any $u \in V$, $(-k)u = k(-u) = -ku$.

Examples of Vector Spaces

(*a*) **Space K^n**: Let K be an arbitrary field. The notation K^n is frequently used to denote the set of all n-tuples of elements in K. Here K^n is a vector space over K using the following operations:

Vector Addition:

$$(a_1, a_2, \ldots, a_n) + (b_1, b_2, \ldots, b_n) = (a_1 + b_1, a_2 + b_2, \ldots, a_n + b_n)$$

Scalar Multiplication:

$$k(a_1, a_2, \dots, a_n) = (ka_1, ka_2, \dots, ka_n)$$

The zero vector in K^n is the n-tuple of zeros, $0 = (0, 0, \dots, 0)$ and the negative of a vector is defined by

$$-(a_1, a_2, \dots, a_n) = (-a_1, -a_2, \dots, -a_n)$$

Observe that these are the same as the operations defined for $\mathbf{R}^n$ in Chapter 1 which we now regard as stating that $\mathbf{R}^n$ with the operations defined there is a vector space over $\mathbf{R}$.

(*b*) **Polynomial Space $\boldsymbol{P}(t)$**: Let $\boldsymbol{P}(t)$ denote the set of all real polynomials of the form

$$p(t) = a_0 + a_1 t + a_2 t^2 + \dots + a_s t^s \qquad (s = 1, 2, \dots)$$

where the coefficients a_i belong to a field K. Then $\boldsymbol{P}(t)$ is a vector space over K using the following operations:
Vector Addition: Here $p(t) + q(t)$ in $\boldsymbol{P}(t)$ is the usual operation of addition of polynomials.
Scalar Multiplication: Here $kp(t)$ in $\boldsymbol{P}(t)$ is the usual operation of the product of a scalar k and a polynomial $p(t)$.
The zero polynomial 0 is the zero vector in $\boldsymbol{P}(t)$.

(*c*) **Polynomial Space $\boldsymbol{P}_n(t)$**: Let $\boldsymbol{P}_n(t)$ denote the set of all real polynomials $p(t)$ over a field K, where the degree of $p(t)$ is less than or equal to n, that is

$$p(t) = a_0 + a_1 t + a_2 t^2 + \dots + a_s t^s$$

where $s \le n$. Then $\boldsymbol{P}_n(t)$ is a vector space over K with respect to the usual operations of addition of polynomials and of multiplication of a polynomial by a constant (just like the vector space $\boldsymbol{P}(t)$ above). We include the zero polynomial 0 as an element of $\boldsymbol{P}_n(t)$, even though its degree is undefined.

(*d*) **Matrix Space $\mathbf{M}_{m,n}$**: The notation $\mathbf{M}_{m,n}$, or simply $\mathbf{M}$, will be used to denote the set of all $m \times n$ matrices with entries in a field K. Then $\mathbf{M}_{m,n}$ is a vector space over K with respect to the usual operations of matrix addition and scalar multiplication of matrices, as indicated by Theorem 2.1.

Linear Combinations; Spanning Sets

Let V be a vector space over a field K. A vector v in V is a *linear combination* of vectors $u_1, u_2, \ldots, u_m$ in V if there exist scalars $a_1, a_2, \ldots, a_m$ in K such that $v = a_1u_1 + a_2u_2 + \ldots + a_mu_m$. Alternatively, v is a linear combination of $u_1, u_2, \ldots, u_m$ if there is a solution to the vector equation $v = x_1u_1 + x_2u_2 + \ldots + x_mu_m$ where $x_1, x_2, \ldots, x_m$ are unknown scalars.

Example 4.1. (Linear Combinations in R^3) Suppose we want to express $v = (3, 7, -4)$ in R^3 as a linear combination of the vectors

$$u_1 = (1, 2, 3), \quad u_2 = (2, 3, 7), \quad u_3 = (3, 5, 6).$$

We seek scalars x, y, z such that $v = xu_1 + yu_2 + zu_3$; that is,

$$\begin{bmatrix} 3 \\ 7 \\ -4 \end{bmatrix} = x\begin{bmatrix} 1 \\ 2 \\ 3 \end{bmatrix} + y\begin{bmatrix} 2 \\ 3 \\ 7 \end{bmatrix} + z\begin{bmatrix} 3 \\ 5 \\ 6 \end{bmatrix} \quad \text{or} \quad \begin{aligned} x+2y+3z&=3 \\ 2x+3y+5z&=7 \\ 3x+7y+6z&=-4 \end{aligned}$$

Reducing the system to echelon form yields

$$\begin{aligned} x+2y+3z&=3 \\ -y-z&=1 \\ y-3z&=-13 \end{aligned} \qquad \textit{then} \qquad \begin{aligned} x+2y+3z&=3 \\ -y-z&=1 \\ -4z&=-12 \end{aligned}$$

Back-substitution yields the solution $x = 2$, $y = -4$, $z = 3$. Thus

$$v = 2u_1 - 4u_2 + 3u_3.$$

Generally speaking, the question of expressing a given vector v in K^n as a linear combination of vectors $u_1, u_2, \ldots, u_m$ in K^n is equivalent to solving a system $AX = B$ of linear equations, where v is the column B of constants, and the u's are the columns of the coefficient matrix A. Such a system may have a unique solution (as above), many solutions, or no solution. The last case – no solution – means that v cannot be written as a linear combination of the u's.

Example 4.2. (Linear combinations in $\boldsymbol{P}(t)$) Suppose we want to express the polynomial $v = 3t^2 + 5t - 5$ as a linear combination of the polynomials $p_1 = t^2 + 2t + 1$, $p_2 = 2t^2 + 5t + 4$, $p_3 = t^2 + 3t + 6$.

We seek scalars x, y, z such that $v = xp_1 + yp_2 + zp_3$; that is,

$$3t^2 + 5t - 5 = x(t^2 + 2t + 1) + y(2t^2 + 5t + 4) + z(t^2 + 3t + 6) \qquad (*)$$

There are two ways to proceed from here.

(1) Expand the right-hand side of (*) obtaining:

$$\begin{aligned} 3t^2 + 5t - 5 &= xt^2 + 2xt + x + 2yt^2 + 5yt + 4y + zt^2 + 3zt + 6z \\ &= (x + 2y + z)t^2 + (2x + 5y + 3z)t + (x + 4y + 6z) \end{aligned}$$

Set coefficients of the same powers of t equal to each other, and reduce the system to echelon form:

$$\begin{aligned} x + 2y + z &= 3 \\ 2x + 5y + 3z &= 5 \\ x + 4y + 6z &= -5 \end{aligned} \quad \text{or} \quad \begin{aligned} x + 2y + z &= 3 \\ y + z &= -1 \\ 2y + 5z &= -8 \end{aligned} \quad \text{or} \quad \begin{aligned} x + 2y + z &= 3 \\ y + z &= -1 \\ 3z &= -6 \end{aligned}$$

The system is in triangular form and has a solution. Back-substitution yields the solution $x = 3$, $y = 1$, $z = -2$. Thus

$$v = 3p_1 + p_2 - 2p_3.$$

(2) The equation (*) is actually an identity in the variable t; that is, the equation holds for any value of t. We can obtain three equations in the unknowns x, y, z by setting t equal to any three values. For example:

$$\begin{aligned} &\text{Set } t = 0 \text{ in } (*) \text{ to obtain:} && x + 4y + 6z = -5 \\ &\text{Set } t = 1 \text{ in } (*) \text{ to obtain:} && 4x + 11y + 10z = 3 \\ &\text{Set } t = -1 \text{ in } (*) \text{ to obtain:} && y + 4z = -7 \end{aligned}$$

Reducing this system to echelon form and solving by back-substitution again yields the solution $x = 3$, $y = 1$, $z = -2$. Thus (again) $v = 3p_1 + p_2 - 2p_3$.

Let V be a vector space over K. Vectors $u_1, u_2, \ldots, u_m$ in V are said to *span* V or to form a *spanning set* of V if every v in V is a linear combination of the vectors $u_1, u_2, \ldots, u_m$, that is, if there exist scalars $a_1, a_2, \ldots, a_m$ in K such that $v = a_1u_1 + a_2u_2 + \cdots + a_mu_m$.

Suppose $u_1, u_2, \ldots, u_m$ span V. Then, for any vector w, the set $w, u_1, u_2, \ldots, u_m$ also spans V.

Suppose $u_1, u_2, \ldots, u_m$ span V and suppose u_k is a linear combination of some of the other u's. Then the u's without u_k also span V.

Suppose $u_1, u_2, \ldots, u_m$ span V and suppose one of the u's is the zero vector. Then the u's without the zero vector also span V.

Example 4.3. Consider the vector space $V = \mathbf{R}^3$.

(*a*) We claim that the following vectors form a spanning set of $\mathbf{R}^3$:

$$e_1 = (1, 0, 0), \qquad e_2 = (0,1,0), \qquad e_3 = (0, 0, 1)$$

Specifically, if $v = (a, b, c)$ is any vector in $\mathbf{R}^3$, then

$$v = ae_1 + be_2 + ce_3.$$

For example, $v = (5, -6, 2) = 5e_1 - 6e_2 + 2e_3$.

(*b*) We claim that the following vectors also form a spanning set of $\mathbf{R}^3$:

$$w_1 = (1, 1, 1), \qquad w_2 = (1, 1, 0), \qquad w_3 = (1, 0, 0)$$

Specifically, if $v = (a, b, c)$ is any vector in $\mathbf{R}^3$, then

$$v = (a,b,c) = cw_1 + (b - c)w_2 + (a - b)w_3.$$

For example, $v = (5, -6, 2) = 2w_1 - 8w_2 + 11w_3$.

(*c*) One can show that $v = (2, 7, 8)$ cannot be written as a linear combination of the vectors

$$u_1 = (1, 2, 3), \qquad u_2 = (1, 3, 5), \qquad u_3 = (1, 5, 9)$$

Accordingly, u_1, u_2, u_3 do not span $\mathbf{R}^3$.

Example 4.4. Consider the vector space $V = \boldsymbol{P}_n(t)$ consisting of all polynomials of degree $\leq n$.

(*a*) Clearly every polynomial in $\boldsymbol{P}_n(t)$ can be expressed as a linear combination of the $n + 1$ polynomials $1, t, t^2, t^3, \ldots, t^n$
Thus these powers of t (where $1 = t^0$) form a spanning set for $\boldsymbol{P}_n(t)$.

(*b*) One can also show that, for any scalar c, the following $n + 1$ powers of $t - c$,

$$1, t - c, (t - c)^2, (t - c)^3, \ldots, (t - c)^n$$

(where $(t - c)^0 = 1$) also form a spanning set for $\boldsymbol{P}_n(t)$.

Example 4.5. Consider the vector space $\mathbf{M} = \mathbf{M}_{2,2}$ consisting of all 2×2 matrices, and consider the following four matrices in $\mathbf{M}$:

$$E_{11} = \begin{bmatrix} 1 & 0 \\ 0 & 0 \end{bmatrix} \quad E_{12} = \begin{bmatrix} 0 & 1 \\ 0 & 0 \end{bmatrix} \quad E_{21} = \begin{bmatrix} 0 & 0 \\ 1 & 0 \end{bmatrix} \quad E_{22} = \begin{bmatrix} 0 & 0 \\ 0 & 1 \end{bmatrix}$$

Then clearly any matrix A in $\mathbf{M}$ can be written as a linear combination of the four matrices. For example,

$$A = \begin{bmatrix} 5 & -6 \\ 7 & 8 \end{bmatrix} = 5E_{11} - 6E_{12} + 7E_{21} + 8E_{22}$$

Accordingly, the four matrices $E_{11}, E_{12}, E_{21}, E_{22}$ span $\mathbf{M}$.

Subspaces

Let V be a vector space over a field K and let W be a subset of V. Then W is a *subspace* of V if W is itself a vector space over K with respect to the operations of vector addition and scalar multiplication on the vector space V.

The way in which one shows that any set W is a vector space is to show that W satisfies the eight axioms of a vector space. However, if W is a subset of a vector space V, then some of the axioms automatically hold in W, since they already hold in V. Simple criteria for identifying subspaces follow.

Theorem 4.2: Suppose W is a subset of a vector space V. Then W is a subspace of V if the following two conditions hold:

(*a*) The zero vector 0 belongs to W.
(*b*) For every $u, v \in W, k \in K$:
(i) The sum $u + v \in W$, (ii) The multiple $ku \in W$.

Property (i) in (*b*) states that W is *closed under vector addition*, and property (ii) in (*b*) states that W is *closed under scalar multiplication*. Both properties may be combined into the following equivalent single statement:

(b') For every $u, v \in W$, $a, b \in K$, the linear combination $au + bv \in W$.

Now let V be any vector space. Then V automatically contains two subspaces, the set $\{0\}$ consisting of the zero vector alone and the whole space V itself. These are sometimes called the *trivial* subspaces of V. Examples of nontrivial subspaces follow.

Example 4.6. Consider the vector space $V = \boldsymbol{R}^3$.
Let U consist of all vectors in $\boldsymbol{R}^3$ whose entries are equal; that is,

$$U = \{(a, b, c): a = b = c\}.$$

For example, (1, 1, 1), (−3, −3, −3), (7, 7, 7), and (−2, −2, −2) are vectors in U. Geometrically, U is the line through the origin O and the point (1, 1, 1). Clearly $0 = (0, 0, 0)$ belongs to U, since all entries in 0 are equal Further, suppose u and v are arbitrary vectors in U, say $u = (a, a, a)$ and $v = (b, b, b)$. Then, for any scalar $k \in \boldsymbol{R}$, the following are also vectors in U:

$$u + v = (a + b, a + b, a + b) \quad \text{and} \quad ku = (ka, ka, ka).$$

Thus U is a subspace of $\boldsymbol{R}^3$.

Example 4.7.

(*a*) Let $V = \mathbf{M}_{n,n}$, the vector space of $n \times n$ matrices. Let W_1 be the subset of all (upper) triangular matrices and let W_2 be the subset of all symmetric matrices. Then W_1 is a subspace of V, since W_1 contains the zero matrix 0 and W_1 is closed under matrix addition and scalar multiplication, that is, the sum and scalar multiple of such triangular matrices are also triangular. Similarly, W_2 is a subspace of V.

(*b*) Let $V = \mathbf{P}(t)$, the vector space $\mathbf{P}(t)$ of polynomials. Then the space $\mathbf{P}_n(t)$ of polynomials of degree at most n may be viewed as a subspace of $\mathbf{P}(t)$. Let $\mathbf{Q}(t)$ be the collection of polynomials with only even powers of t. For example, the following are polynomials in $\mathbf{Q}(t)$:

$$p_1 = 3 + 4t^2 - 5t^6 \text{ and } p_2 = 6 - 7t^4 + 9t^6 + 3t^{12}$$

(We assume that any constant $k = kt^0$ is an even power of t.) Then $\mathbf{Q}(t)$ is a subspace of $\mathbf{P}(t)$.

Let U and W be subspaces of a vector space V. We show that the intersection $U \cap W$ is also a subspace of V. Clearly, $0 \in U$ and $0 \in W$, since U and W are subspaces; whence $0 \in U \cap W$. Now suppose u and v belong to the intersection $U \cap W$. Then $u, v \in U$ and $u, v \in W$. Further, since U and W are subspaces, for any scalars $a, b \in K$, $au + bv \in U$ and $au + bv \in W$. Thus $au + bv \in U \cap W$. Therefore $U \cap W$ is a subspace of V.

Theorem 4.3: The intersection of any number of subspaces of a vector space V is a subspace of V.

Consider a system $AX = B$ of linear equations in n unknowns. Then every solution u may be viewed as a vector in K^n. Thus the solution set of such a system is a subset of K^n. Now suppose the system is homogeneous, that is, suppose the system has the form $AX = 0$. Let W be its solution set. Since $A0 = 0$, the zero vector $0 \in W$. Moreover, suppose u and v belong to W. Then u and v are solutions of $AX = 0$, or, in other words, $Au = 0$ and $Av = 0$. Therefore, for any scalars a and b, we have $A(au + bv) = aAu + bAv = a0 + b0 = 0 + 0 = 0$. Thus $au + bv$ belongs to W, since it is a solution of $AX = 0$. Accordingly, W is a subspace of K^n.

Theorem 4.4: The solution set W of a homogeneous system $AX = 0$ in n unknowns is a subspace of K^n.

We emphasize that the solution set of a nonhomogeneous system $AX = B$ is not a subspace of K^n. In fact, the zero vector 0 does not belong to its solution set.

Linear Span; Row Space of a Matrix

Suppose $u_1, u_2, \dots, u_m$ are any vectors in a vector space V. Recall that any vector of the form $a_1u_1 + a_2u_2 + \dots + a_m u_m$, where the a_i are scalars, is called a *linear combination* of $u_1, u_2, \dots, u_m$. The collection of all such linear combinations denoted by span$(u_1, u_2, \dots, u_m)$ or span(u_i) is called the *linear span* of $u_1, u_2, \dots, u_m$.

Clearly the zero vector 0 belongs to span (u_i), since

$$0 = 0u_1 + 0u_2 + \dots + 0u_m$$

Furthermore, suppose v and v' belong to span (u_i), say

$$v = a_1u_1 + a_2u_2 + \dots + a_m u_m \text{ and } v' = b_1u_1 + b_2u_2 + \dots + b_m u_m$$

and

$$kv = ka_1u_1 + ka_2u_2 + \dots + ka_m u_m$$

Thus $v + v'$ and kv also belong to span(u_i). Accordingly, span(u_i) is a subspace of V.

More generally, for any subset S of V, span(S) consists of all linear combinations of vectors in S or, when $S = \varnothing$, span$(S) = \{0\}$. Thus, in particular, S is a spanning set of span(S).

Theorem 4.5: Let S be a subset of a vector space V.

(i) Then span(S) is a subspace of V that contains S.

(ii) If W is a subspace of V containing S, then span$(S) \subseteq W$.

Condition (ii) in Theorem 4.5 may be interpreted as saying that span(S) is the "smallest" subspace of V containing S.

Example 4.8. Consider the vector space $V = \mathrm{R}^3$. Let u be any nonzero vector in R^3. Then span(u) consists of all scalar multiples of u.

Let $A = [a_{ij}]$ be an arbitrary $m \times n$ matrix over a field K. The rows of A, $R_1 = (a_{11}, a_{12}, \dots, a_{1n})$, $R_2 = (a_{21}, a_{22}, \dots, a_{2n})$, $\dots$, $R_m = (a_{m1}, a_{m2}, \dots, a_{mn})$, may be viewed as vectors in K^n; hence they span a subspace of K^n called the *row space* of A and denoted by rowsp(A). That is, rowsp(A) = span$(R_1, R_2, \dots, R_m)$.

Analogously, the columns of A may be viewed as vectors in K^m

called the *column space* of A and denoted by colsp(A). Observe that colsp(A) = rowsp(A^T).

Recall that matrices A and B are row equivalent, written $A \sim B$, if B can be obtained from A by a sequence of elementary row operations. Now suppose M is the matrix obtained by applying one of the elementary row operations on a matrix A. Then each row of M is a row of A or a linear combination of rows of A. Hence the row space of M is contained in the row space of A. On the other hand, we can apply the inverse elementary row operations on M to obtain A; hence the row space of A is contained in the row space of M. Accordingly, A and M have the same row space. This will be true each time we apply an elementary row operation. Thus we have the following two theorems:

Theorem 4.6: Row equivalent matrices have the same row space.

Theorem 4.7: Suppose A and B are row canonical matrices. Then A and B have the same row space if and only if they have the same nonzero rows.

Example 4.9. Consider the following two sets of vectors in R^4:

$$u_1 = (1, 2, -1, 3),\ u_2 = (2, 4, 1, -2),\ u_3 = (3, 6, 3, -7)$$
$$w_1 = (1, 2, -4, 11),\ w_2 = (2, 4, -5, 14)$$

Let $U = \mathrm{span}(u_i)$ and $W = \mathrm{span}(w_i)$.

There are two ways to show that $U = W$.

(*a*) Show that each u_i is a linear combination of w_1 and w_2, and show that each w_i is a linear combination of u_1, u_2, u_3. Observe that we have to show that six systems of linear equations are consistent.

(*b*) Form the matrix A whose rows are u_1, u_2, u_3 and row reduce A to row canonical form, and form the matrix B whose rows are w_1 and w_2 and row reduce B to row canonical form:

$$A = \begin{bmatrix} 1 & 2 & -1 & 3 \\ 2 & 4 & 1 & -2 \\ 3 & 6 & 3 & -7 \end{bmatrix} \sim \begin{bmatrix} 1 & 2 & -1 & 3 \\ 0 & 0 & 3 & -8 \\ 0 & 0 & 6 & -16 \end{bmatrix} \sim \begin{bmatrix} 1 & 2 & 0 & \frac{1}{3} \\ 0 & 0 & 1 & \frac{-8}{3} \\ 0 & 0 & 0 & 0 \end{bmatrix}$$

$$B = \begin{bmatrix} 1 & 2 & -4 & 11 \\ 2 & 4 & -5 & 14 \end{bmatrix} \sim \begin{bmatrix} 1 & 2 & -4 & 11 \\ 0 & 0 & 3 & -8 \end{bmatrix} \sim \begin{bmatrix} 1 & 2 & 0 & \frac{1}{3} \\ 0 & 0 & 1 & \frac{-8}{3} \end{bmatrix}$$

Since the nonzero rows of the matrices in row canonical form are identical, the row spaces of A and B are equal. Therefore $U = W$.

Clearly, the method in (*b*) is more efficient than the method in (*a*).

Linear Dependence and Independence

We say that the vectors $v_1, v_2, \ldots, v_m$ in V are *linearly dependent* if there exist scalars $a_1, a_2, \ldots, a_m$ in K, not all of them 0, such that

$$a_1v_1 + a_2v_2 + \ldots + a_mv_m = 0.$$

Otherwise, we say that the vectors are *linearly independent.*

The above definition may be restated as follows. Consider the vector equation

$$x_1v_1 + x_2v_2 + \ldots + x_mv_m = 0 \qquad (*)$$

where the x's are unknown scalars. This equation always has the *zero solution* $x_1 = 0, x_2 = 0, \ldots, x_m = 0$. Suppose this is the only solution, that is, suppose we can show:

$$x_1v_1 + x_2v_2 + \ldots + x_mv_m = 0 \text{ implies } x_1 = 0, x_2 = 0, \ldots, x_m = 0.$$

Then the vectors $v_1, v_2, \ldots, v_m$ are linearly independent. On the other hand, suppose the equation (*) has a nonzero solution; then the vectors are linearly dependent.

A set $S = \{v_1, v_2, \ldots, v_m\}$ of vectors in V is linearly dependent or independent according as the vectors $v_1, v_2, \ldots, v_m$ are linearly dependent or independent.

The following properties come directly from the above definition.

(*a*) Suppose 0 is one of the vectors $v_1, v_2, \ldots, v_m$, say $v_1 = 0$. Then the vectors must be linearly dependent, since we have the following linear combination where the coefficient of $v_1 \neq 0$:

$$1v_1 + 0v_2 + \ldots + 0v_m = 1 \cdot 0 + 0 \ldots + 0 = 0$$

(*b*) Suppose v is a nonzero vector. Then v, by itself, is linearly independent, since $kv = 0$, $v \neq 0$ implies $k = 0$.

(*c*) Suppose two of the vectors $v_1, v_2, \ldots, v_m$ are equal or one is a scalar multiple of the other, say $v_1 = kv_2$. Then the vectors must be linearly dependent, since we have the following linear combination where the coefficient of $v_1 \neq 0$:

$$v_1 - kv_2 + 0v_3 + \ldots + 0v_m = 0$$

(*d*) Two vectors v_1, v_2 are linearly dependent if and only if one of them is a multiple of the other.

(*e*) If a set S of vectors is linearly independent, then any subset of S is linearly independent. Alternatively, if S contains a linearly dependent subset, then S is linearly dependent.

Example 4.10.

(*a*) Let $u = (1, 1, 0)$, $v = (1, 3, 2)$, $w = (4, 9, 5)$. Then u, v, w are linearly dependent, since

$$3u + 5v - 2w = 3(1, 1, 0) + 5(1, 3, 2) - 2(4, 9, 5) = (0, 0, 0) = 0.$$

(*b*) We show that the vectors $u = (1, 2, 3)$, $v = (2, 5, 7)$, $w = (1, 3, 5)$ are linearly independent. We form the vector equation $xu + yv + zw = 0$ where x, y, z are unknown scalars. This yields

$$x\begin{bmatrix}1\\2\\3\end{bmatrix} + y\begin{bmatrix}2\\5\\7\end{bmatrix} + z\begin{bmatrix}1\\3\\5\end{bmatrix} = \begin{bmatrix}0\\0\\0\end{bmatrix} \quad \text{or} \quad \begin{aligned} x + 2y + z &= 0 \\ 2x + 5y + 3y &= 0 \\ 3x + 7y + 5z &= 0 \end{aligned} \quad \text{or} \quad \begin{aligned} x + 2y + z &= 0 \\ y + z &= 0 \\ z &= 0 \end{aligned}$$

Back-substitution yields $x = 0$, $y = 0$, $z = 0$. We have shown that

$$xu + yv + zw = 0 \text{ implies } x = 0, y = 0, z = 0.$$

The notions of linear dependence and linear combinations are closely related. Specifically, for more than one vector, we show that the vectors $v_1, v_2, \ldots, v_m$ are linearly dependent if and only if one of them is a linear combination of the others,

$$v_i = a_1 v_1 + \ldots + a_{i-1} v_{i-1} + a_{i+1} v_{i+1} + \ldots + a_m v_m$$

Then by adding $-v_i$ to both sides, we obtain

$$a_1 v_1 + \ldots + a_{i-1} v_{i-1} - v_i + a_{i+1} v_{i+1} + \ldots + a_m v_m = 0.$$

Where the coefficient of v_i is not 0. Hence the vectors are linearly dependent. Conversely, suppose the vectors are linearly dependent, say, $b_1 v_1 + \ldots + b_j v_j + \ldots + b_m v_m = 0$, where $b_j \neq 0$. Then we can solve for v_j obtaining

$$v_j = b_j^{-1} b_1 v_1 - \ldots - b_j^{-1} b_{j-1} v_{j-1} - b_j^{-1} b_{j+1} v_{j+1} - \ldots - b_j^{-1} b_m v_m$$

and so v_j is a linear combination of the other vectors.

Lemma 4.8: Suppose two or more nonzero vectors $v_1, v_2, \ldots, v_m$ are linearly dependent. Then one of the vectors is a linear combination of the preceding vectors, that is, there exists $k > 1$ such that

$$v_k = c_1 v_1 + c_2 v_2 + \ldots + c_{k-1} v_{k-1}$$

Consider the following echelon matrix A, whose pivots have been shaded:

$$A = \begin{bmatrix} 0 & 2 & 3 & 4 & 5 & 6 & 7 \\ 0 & 0 & 4 & 3 & 2 & 3 & 4 \\ 0 & 0 & 0 & 0 & 7 & 8 & 9 \\ 0 & 0 & 0 & 0 & 0 & 6 & 7 \\ 0 & 0 & 0 & 0 & 0 & 0 & 0 \end{bmatrix}$$

Observe the rows R_2, R_3, R_4 have 0's in the second column below the nonzero pivot in R_1, and hence any linear combination of R_2, R_3, R_4 must have 0 as its second entry. Thus R_1 cannot be a linear combination of the rows below it. Similarly, the rows R_3 and R_4 have 0's in the third column below the nonzero pivot in R_2 and hence R_2 cannot be a linear combination of the rows below it. Finally, R_3 cannot be a multiple of R_4 since R_4 has a 0 in the fifth column below the nonzero pivot in R_3. Viewing the nonzero rows from the bottom up, R_4, R_3, R_2, R_1, no row is a linear combination of the preceding rows. Thus the rows are linearly independent by Lemma 4.8.

The argument used with the above echelon matrix A can be used for the nonzero rows of any echelon matrix. Thus we have the following very useful result.

Theorem 4.9: The nonzero rows of a matrix in echelon form are linearly independent.

Basis and Dimension

A set $S = \{u_1, u_2, \ldots, u_n\}$ of vectors is a basis of V if it has the following two properties: (1) S is linearly independent. (2) S spans V.

Alternately, a set $S = \{u_1, u_2, \ldots, u_n\}$ of vectors is a basis of V if every $v \in V$ can be written uniquely as a linear combination of the basis vectors.

Theorem 4.10: Let V be a vector space such that one basis has m elements and another basis has n elements. Then $m = n$.

A vector space V is said to be of *finite dimension n* or *n-dimensional*, written $\dim V = n$ if V has a basis with n elements. Theorem 4.10 tells us that all bases of V have the same number of elements, so this definition is well-defined.

The vector space $\{0\}$ is defined to have dimension 0.

Suppose a vector space V does not have a finite basis. Then V is said to be of *infinite dimension* or to be *infinite-dimensional.*

The above fundamental Theorem 4.10 is a consequence of the following "replacement lemma."

Lemma 4.11: Suppose $\{v_1, v_2, \ldots, v_n\}$ spans V, and suppose $\{w_1, w_2, \ldots, w_m\}$ is linearly independent. Then $m \leq n$, and V is spanned by a set of the form $\{w_1, w_2, \ldots, w_m, v_{i_1}, v_{i_2}, \ldots, v_{i_{n-m}}\}$. Thus, in particular, $n + 1$ or more vectors in V are linearly dependent.

Observe in the above lemma that we have replaced m of the vectors in the spanning set of V by the m independent vectors and still retained a spanning set.

Examples of Bases

This subsection presents important examples of bases of some of the main vector spaces appearing in this text.

(*a*) **Vector space K^n**: Consider the following n vectors in K^n:

$$e_1 = (1, 0, 0, 0, \ldots 0, 0),\ e_2 = (0, 1, 0, 0, \ldots 0, 0), \ldots,\\ e_n = (0, 0, 0, 0, \ldots, 0, 1)$$

These vectors are linearly independent. (For example, they form a matrix in echelon form.) Furthermore, any vector $u = (a_1, a_2, \ldots, a_n)$ in K^n can be written as a linear combination of the above vectors. Specifically, $v = a_1e_1 + a_2e_2 + \ldots + a_ne_n$. Accordingly, the vectors form a basis of K^n called the usual or standard basis of K^n. Thus (as one might expect) K^n has dimension n. In particular, any other basis of K^n has n elements.

(*b*) **Vector space $\mathbf{M} = \mathbf{M}_{r,s}$ of all $r \times s$ matrices**: The following six matrices form a basis of the vector space $\mathbf{M}_{2,3}$ of all 2×3 matrices over K:

$$\begin{bmatrix}1 & 0 & 0\\0 & 0 & 0\end{bmatrix}, \begin{bmatrix}0 & 1 & 0\\0 & 0 & 0\end{bmatrix}, \begin{bmatrix}0 & 0 & 1\\0 & 0 & 0\end{bmatrix}, \begin{bmatrix}0 & 0 & 0\\1 & 0 & 0\end{bmatrix}, \begin{bmatrix}0 & 0 & 0\\0 & 1 & 0\end{bmatrix}, \begin{bmatrix}0 & 0 & 0\\0 & 0 & 1\end{bmatrix}$$

More generally, in the vector space $\mathbf{M} = \mathbf{M}_{r,s}$ of all $r \times s$ matrices, let E_{ij} be the matrix with ij-entry 1 and 0's elsewhere. Then all such matrices form a basis of $\mathbf{M}_{r,s}$ called the *usual* or *standard* basis of $\mathbf{M}_{r,s}$. Accordingly, $\dim \mathbf{M}_{r,s} = rs$.

(*c*) **Vector space $\mathbf{P}_n(t)$ of all polynomials of degree $\leq n$**: The set

$S = \{1, t, t^2, t^3, \ldots, t^n\}$ of $n + 1$ polynomials is a basis of $\mathbf{P}_n(t)$.

Specifically, any polynomial $f(t)$ of degree $\leq n$ can be expressed as a linear combination of these powers of t, and one can show that these polynomials are linearly independent. Therefore,

$$\dim \mathbf{P}_n(t) = n + 1.$$

The following three theorems will be used frequently.

Theorem 4.12: Let V be a vector space of finite dimension n. Then:

(i) Any $n + 1$ or more vectors in V are linearly dependent.

(ii) Any linearly independent set $S = \{u_1, u_2, \ldots, u_n\}$ with n elements is a basis of V.

(iii) Any spanning set $T = \{v_1, v_2, \ldots, v_n\}$ of V with n elements is a basis of V.

Theorem 4.13: Suppose S spans a vector space V. Then:

(i) Any maximum number of linearly independent vectors in S form a basis of V.

(ii) Suppose one deletes from S every vector that is a linear combination of preceding vectors in S. Then the remaining vectors form a basis of V.

Theorem 4.14: Let V be a vector space of finite dimension and let $S = \{u_1, u_2, \ldots, u_n\}$ be a set of linearly independent vectors in V. Then S is part of a basis of V; that is, S may be extended to a basis of V.

Example 4.11.

(*a*) The following four vectors in $\mathbf{R}^4$ form a matrix in echelon form:

$$(1, 1, 1, 1), (0, 1, 1, 1), (0, 0, 1, 1), (0, 0, 0, 1)$$

Thus the vectors are linearly independent, and, since $\dim \mathbf{R}^4 = 4$, the vectors form a basis of $\mathbf{R}^4$.

(*b*) The following $n + 1$ polynomials in $\mathbf{P}_n(t)$ are of increasing degree:

$$1, t - 1, (t - 1)^2, \ldots, (t - 1)^n$$

Therefore no polynomial is a linear combination of preceding polynomials; hence the polynomials are linearly independent. Furthermore, they form a basis of $\mathbf{P}_n(t)$, since $\dim \mathbf{P}_n(t) = n + 1$.

(*c*) Consider any four vectors in $\mathbf{R}^3$; say

$$(1, 2, 3), (-2, 3, 4), (5, -2, 7), (13, 3, -8)$$

By Theorem 4.12(i), the four vectors must be linearly dependent, since they come from the 3-dimensional vector space $\mathbf{R}^3$.

The following theorem gives the basic relationship between the dimension of a vector space and the dimension of a subspace.

Theorem 4.15: Let W be a subspace of an n-dimensional vector space V. Then $\dim W \leq n$. In particular, if $\dim W = n$, then $W = V$.

Example 4.12. Let W be a subspace of the real space $\mathbf{R}^3$. Note that $\dim \mathbf{R}^3 = 3$. Theorem 4.15 tells us that the dimension of W can only be 0, 1, 2, or 3. Therefore the following cases apply:

(*a*) $\dim W = 0$, then $W = \{0\}$, a point.
(*b*) $\dim W = 1$, then W is a line through the origin 0.
(*c*) $\dim W = 2$, then W is a plane through the origin 0.
(*d*) $\dim W = 3$, then W is the entire space $\mathbf{R}^3$.

Rank of a Matrix

The *rank* of a matrix A, written $\text{rank}(A)$, is equal to the maximum number of linearly independent rows of A or, equivalently, the dimension of the row space of A.

Theorem 4.16: The maximum number of linearly independent rows of any matrix A is equal to the maximum number of linearly independent columns of A. Thus the dimension of the row space of A is equal to the dimension of the column space of A.

Accordingly, one could restate the above definition of the rank of A using column instead of row.

An echelon form of any matrix A gives us the solution to certain problems about A itself. Specifically, let A and B be the following matrices, where the echelon matrix B (whose pivots are shaded) is an echelon form of A:

$$A = \begin{bmatrix} 1 & 2 & 1 & 3 & 1 & 2 \\ 2 & 5 & 5 & 6 & 4 & 5 \\ 3 & 7 & 6 & 11 & 6 & 9 \\ 1 & 5 & 10 & 7 & 9 & 9 \\ 2 & 6 & 8 & 12 & 9 & 12 \end{bmatrix} \quad \text{and} \quad B = \begin{bmatrix} 1 & 2 & 1 & 3 & 1 & 2 \\ 0 & 1 & 3 & 1 & 2 & 1 \\ 0 & 0 & 0 & 1 & 1 & 2 \\ 0 & 0 & 0 & 0 & 0 & 0 \\ 0 & 0 & 0 & 0 & 0 & 0 \end{bmatrix}$$

We solve the following four problems about the matrix A, where C_1, C_2, …, C_6 denote its columns:

(*a*) Find a basis of the row space of A.
(*b*) Find each column C_k of A that is a linear combination of preceding columns of A.
(*c*) Find a basis of the column space of A.
(*d*) Find the rank of A.

(*a*) We are given that A and B are row equivalent, so they have the same row space. Moreover, B is in echelon form, so its nonzero rows are linearly independent and hence form a basis of the row space of B. Thus they also form a basis of the row space of A. That is,

basis of rowsp(A): (1, 2, 1, 3, 1, 2), (0, 1, 3, 0, 2, 1), (0, 0, 0, 1, 1, 2)

(*b*) Let $M_k = [C_1, C_2, \ldots, C_k]$, the submatrix of A consisting of the first k columns of A. Then M_{k-1} and M_k are, respectively, the coefficient matrix and augmented matrix of the vector equation

$$x_1C_1 + x_2C_2 + \ldots + x_{k-1}C_{k-1} = C_k$$

Theorem 3.8 tells us that the system has a solution, or, equivalently; C_k is a linear combination of the preceding columns of A if and only if $\text{rank}(M_k) = \text{rank}(M_{k-1})$, where $\text{rank}(M_k)$ means the number of pivots in an echelon form of M_k. Now the matrix formed by the first k columns of the echelon matrix B is also an echelon form of M_k. Accordingly, $\text{rank}(M_2) = \text{rank}(M_3) = 2$ and $\text{rank}(M_4) = \text{rank}(M_5) = \text{rank}(M_6) = 3$. Thus C_3, C_5, C_6 are each a linear combination of the preceding columns of A.

(*c*) The fact that the remaining columns C_1, C_2, C_4 are not linear combinations of their respective preceding columns also tells us that they are linearly independent. Thus they form a basis of the column space of A. That is

basis of colsp(A): $[1, 2, 3, 1, 2]^T$, $[2, 5, 7, 5, 6]^T$, $[3, 6, 11, 8, 11]^T$

Observe that C_1, C_2, C_4 may also be characterized as those columns of A that contain the pivots in any echelon form of A.

(*d*) Here we see that three possible definitions of the rank of A yield the same value.

(i) There are three pivots in B, which is an echelon form of A.

(ii) The three pivots in B correspond to the nonzero rows of B, which form a basis of the row space of A.

(iii) The three pivots in B correspond to the columns of A, which form a basis of the column space of A.

Thus $\text{rank}(A) = 3$.

Frequently, we are given a list $S = \{u_1, u_2, \ldots, u_r\}$ of vectors in K^n and we want to find a basis for the subspace W of K^n spanned by the given vectors, that is, a basis of $W = \text{span}(S) = \text{span}(u_1, u_2, \ldots, u_r)$. The following two algorithms, which are essentially described in the above subsection, find such a basis (and hence the dimension) of W.

Algorithm 4.1 (Row space algorithm)

Step 1. Form the matrix M whose *rows* are the given vectors.
Step 2. Row reduce M to echelon form.
Step 3. Output the nonzero rows of the echelon matrix.

Sometimes we want to find a basis that only comes from the original given vectors. The next algorithm accomplishes this task.

Algorithm 4.2 (Casting-out algorithm)

Step 1. Form the matrix M whose *columns* are the given vectors.
Step 2. Row reduce M to echelon form.
Step 3. For each column C_k in the echelon matrix without a pivot, delete (cast out) the vector u_k from the list S of given vectors.
Step 4. Output the remaining vectors in S (which correspond to columns with pivots).

Example 4.13. Let W be the subspace of $\mathbf{R}^5$ spanned by the following vectors:

$$u_1 = (1, 2, 1, 3, 2), \quad u_2 = (1, 3, 3, 5, 3), \quad u_3 = (3, 8, 7, 13, 8)$$
$$u_4 = (1, 4, 6, 9, 7), \quad u_5 = (5, 13, 13, 25, 19)$$

Find a basis of W consisting of the original given vectors, and find $\dim W$.

Form the matrix M whose columns are the given vectors and reduce M to echelon form:

$$M = \begin{bmatrix} 1 & 1 & 3 & 1 & 5 \\ 2 & 3 & 8 & 4 & 13 \\ 1 & 3 & 7 & 6 & 13 \\ 3 & 5 & 13 & 9 & 25 \\ 2 & 3 & 8 & 7 & 19 \end{bmatrix} \sim \begin{bmatrix} 1 & 1 & 3 & 1 & 5 \\ 0 & 1 & 2 & 2 & 3 \\ 0 & 0 & 0 & 1 & 2 \\ 0 & 0 & 0 & 0 & 0 \\ 0 & 0 & 0 & 0 & 0 \end{bmatrix}$$

The pivots in the echelon matrix appear in columns C_1, C_2, C_4. Accordingly, we "cast out" the vectors u_3 and u_5 from the original five vectors. The remaining vectors u_1, u_2, u_4, which correspond to the columns in the echelon matrix with pivots, form a basis of W. Thus, in particular, $\dim W = 3$.

Consider again a homogeneous system $AX = 0$ of linear equations over K with n unknowns. By Theorem 4.4, the solution set W of such a system is a subspace of K^n, and hence W has a dimension. The following theorem holds.

Theorem 4.17: The dimension of the solution space W of a homogeneous system $AX = 0$ is $n - r$, where n is the number of unknowns and r is the rank of the coefficient matrix A.

Sums and Direct Sums

Let U and W be subsets of a vector space V. The sum of U and W, written $U + W$, consists of all sums $u + w$ where $u \in U$ and $w \in W$. That is, $U + W = \{v\colon v = u + w, \text{ where } u \in U \text{ and } w \in W\}$.

Now suppose U and W are subspaces of V. Then one can easily show that $U + W$ is a subspace of V. Recall that $U \cap W$ is also a subspace of V. The following theorem relates the dimensions of these subspaces.

Theorem 4.18: Suppose U and W are finite-dimensional subspaces of a vector space V. Then $U + W$ has finite dimension and

$$\dim(U + W) = \dim U + \dim W - \dim(U \cap W)$$

Example 4.14. Let $V = \mathbf{M}_{2,2}$, the vector space of 2×2 matrices. Let U consist of those matrices whose second row is zero, and let W consist of those matrices whose second column is zero. Then

$$U = \left\{ \begin{bmatrix} a & b \\ 0 & 0 \end{bmatrix} \right\}, \quad W = \left\{ \begin{bmatrix} a & 0 \\ c & 0 \end{bmatrix} \right\}, \quad \text{and} \quad U + W = \left\{ \begin{bmatrix} a & b \\ c & 0 \end{bmatrix} \right\}, \quad U \cap W = \left\{ \begin{bmatrix} a & 0 \\ 0 & 0 \end{bmatrix} \right\}$$

That is, $U + W$ consists of those matrices whose lower right entry is 0, and $U \cap W$ consists of those matrices whose second row and second column are zero. Note that $\dim U = 2$, $\dim W = 2$, $\dim(U \cap W) = 1$. Also, $\dim(U + W) = 3$, which is expected from Theorem 4.18. That is, $\dim(U + W) = \dim U + \dim W - \dim(U \cap W) = 2 + 2 - 1 = 3$.

The vector space V is said to be the *direct sum* of its subspaces U and W, denoted by $V = U \oplus W$ if every $v \in V$ can be written in one and only one way as $v = u + w$ where $u \in U$ and $w \in W$.

Theorem 4.19: The vector space V is the direct sum of its subspaces U and W if and only if: (i) $V = U + W$, (ii) $U \cap W = \{0\}$.

Example 4.15. Consider the vector space $V = \mathbf{R}^3$.

(*a*) Let U be the xy-plane and let W be the yz-plane; that is,

$$U = \{(a, b, 0): a, b \in \mathbf{R}\} \text{ and } W = \{(0, b, c): b, c \in \mathbf{R}\}.$$

Then $\mathbf{R}^3 = U + W$, since every vector in $\mathbf{R}^3$ is the sum of a vector in U and a vector in W. However, $\mathbf{R}^3$ is not the direct sum of U and W, since such sums are not unique. For example,

$$(3, 5, 7) = (3, 1, 0) + (0, 4, 7)$$

and also

$$(3, 5, 7) = (3, -4, 0) + (0, 9, 7)$$

(*b*) Let U be the xy-plane and let W be the z-axis; that is,

$$U = \{(a, b, 0): a, b \in \mathbf{R}\} \text{ and } W = \{(0, 0, c): c \in \mathbf{R}\}.$$

Now any vector $(a, b, c) \in \mathbf{R}^3$ can be written as the sum of a vector in U and a vector in W in one and only one way:

$$(a, b, c) = (a, b, 0) + (0, 0, c)$$

Accordingly, $\mathbf{R}^3$ is the direct sum of U and W; that is, $\mathbf{R}^3 = U \oplus W$.

The notion of a direct sum is extended to more than one factor in the obvious way. That is, V is the *direct sum* of subspaces $W_1, W_2, \ldots, W_r$, written

$$V = W_1 \oplus W_2 \oplus \ldots \oplus W_r$$

if every vector $v \in V$ can be written in one and only one way as

$$v = w_1 + w_2 + \ldots + w_r \text{ where } w_1 \in W_1, w_2 \in W_2, \ldots, w_r \in W_r.$$

Theorem 4.20: Suppose $V = W_1 \oplus W_2 \oplus \ldots \oplus W_r$. Also, for each k suppose S_k is a linearly independent subset of W_k. Then:

(*a*) The union $S = \cup_k S_k$ is linearly independent in V.

(*b*) If each S_k is a basis of W_k, then $\cup_k S_k$ is a basis of V.

(*c*) $\dim V = \dim W_1 + \dim W_2 + \ldots + \dim W_r$.

Theorem 4.21: Suppose $V = W_1 + W_2 + \ldots + W_r$ and $\dim V = \Sigma_k \dim W_k$. Then $V = W_1 \oplus W_2 \oplus \ldots \oplus W_r$.

Chapter 5
Inner Product Spaces; Orthogonality

IN THIS CHAPTER:

- ✔ *Inner Product Spaces*
- ✔ *Cauchy-Schwarz Inequality*
- ✔ *Orthogonality*
- ✔ *Gram-Schmidt Orthogonalization Process*

Inner Product Spaces

The definition of a vector space V involves an arbitrary field K. Here we will mainly restrict K to be the real field $\mathbf{R}$, in which case V is called a *real vector space*. Also we adopt the previous notation that

u, v, w	are vectors in V
a, b, c, k	are scalars in K

Furthermore, the vector spaces V in this chapter have finite dimension unless otherwise stated or implied.

Let V be a real vector space. Suppose to each pair of vectors $u, v \in V$ there is assigned a real number, denoted by $\langle u, v \rangle$. This function is called a (*real*) *inner product* on V if it satisfies the following axioms:

[I_1] (**Linear Property**): $<au_1 + bu_2, v> = a<u_1, v> + b<u_2, v>$.
[I_2] (**Symmetric Property**): $<u, v> = <v, u>$.
[I_3] (**Positive Definite Property**): $<u, u> \geq 0$; and $<u, u> = 0$ if and only if $u = 0$.

The vector space V with an inner product is called a (*real*) *inner product space.*

By the third axiom [I_3] of an inner product, $<u, u>$ is nonnegative for any vector u. Thus its positive square root exists. We use the notation $\|u\| = \sqrt{<u, u>}$. This nonnegative number is called the *norm* or *length* of u. The relation $\|u\|^2 = <u, u>$ will be used frequently.

If $\|u\| = 1$ or, equivalently, if $<u, u> = 1$, then u is called a *unit vector* and is said to be *normalized.* Every nonzero vector v in V can be multiplied by the reciprocal of its length to obtain the unit vector

$$\hat{v} = \frac{1}{\|v\|} v$$

which is a positive multiple of v. This process is called *normalizing* v.

Examples of Inner Product Spaces

This subsection lists the main examples of inner product spaces used in this text.

(a) *Euclidean n-Space* $\mathbf{R}^n$: Consider the vector space $\mathbf{R}^n$. The *dot product* or *scalar product* in $\mathbf{R}^n$ is defined by $u \cdot v = a_1 b_1 + a_2 b_2 + \ldots + a_n b_n$ where $u = (a_i)$ and $v = (b_i)$. This function defines an inner product on $\mathbf{R}^n$. The norm $\|u\|$ of the vector $u = (a_i)$ in this space is as follows:

$$\|u\| = \sqrt{u \cdot u} = \sqrt{a_1^2 + a_2^2 + \ldots + a_n^2}$$

On the other hand, by the Pythagorean theorem, the distance from the origin O in $\mathbf{R}^3$ to a point $P(a, b, c)$ is given by $\sqrt{a^2 + b^2 + c^2}$. This is precisely the same as the above-defined norm of the vector $v = (a, b, c)$ in $\mathbf{R}^3$. Since the Pythagorean theorem is a consequence of the axioms of Euclidean geometry, the vector space

$\mathbf{R}^n$ with the above inner product and norm is called *Euclidean n-space*. Although there are many ways to define an inner product on $\mathbf{R}^n$, we shall assume this inner product unless otherwise stated or implied. It is called the *usual* (or *standard*) *inner product* on $\mathbf{R}^n$.

Frequently the vectors in $\mathbf{R}^n$ will be represented by column vectors, that is, by $n \times 1$ column matrices. In such a case, the formula $\langle u, v\rangle = u^T v$ defines the usual inner product on $\mathbf{R}^n$.

Example 5.1. Let $u = (1, 3, -4, 2)$, $v = (4, -2, 2, 1)$, $w = (5, -1, -2, 6)$ be vectors in $\mathbf{R}^4$.
By definition,

$$\langle u, w\rangle = 5 - 3 + 8 + 12 = 22 \text{ and } \langle v, w\rangle = 20 + 2 - 4 + 6 = 24.$$

Note that $3u - 2v = (-5, 13, -16, 4)$. Thus

$$\langle 3u - 2v, w\rangle = -25 - 13 + 32 + 24 = 18.$$

(*b*) *Polynomial Space* $\mathbf{P}(t)$: Consider the vector space $\mathbf{P}(t)$ of all polynomials. The following defines an inner product where $f(t)$ and $g(t)$ are polynomials in $\mathbf{P}(t)$:

$$\langle f, g\rangle = \int_a^b f(t)g(t)dt$$

It is called the usual inner product on $\mathbf{P}(t)$.

Example 5.2. Consider $f(t) = 3t - 5$ and $g(t) = t^2$ in the polynomial space $\mathbf{P}(t)$ with inner product $\langle f, g\rangle = \int_0^1 f(t)g(t)dt$.

(*a*) Find $\langle f, g\rangle$.
We have $f(t)g(t) = 3t^3 - 5t^2$. Hence

$$\langle f, g\rangle = \int_0^1 \left(3t^3 - 5t^2\right)dt = \frac{3}{4}t^4 - \frac{5}{3}t^3\Big|_0^1 = \frac{3}{4} - \frac{5}{3} = -\frac{11}{12}$$

(*b*) Find $\|f\|$ and $\|g\|$.

We have $[f(t)]^2 = 9t^2 - 30t + 25$ and $[g(t)]^2 = t^4$. Then

$$\|f\|^2 = \langle f, f \rangle = \int_0^1 \left(9t^2 - 30t + 25\right)dt = 3t^3 - 15t^2 + 25t\Big|_0^1 = 13$$

$$\|g\|^2 = \langle g, g \rangle = \int_0^1 t^4 dt = \frac{1}{5}t^5\Big|_0^1 = \frac{1}{5}$$

Therefore, $\|f\| = \sqrt{13}$ and $\|g\| = \sqrt{\frac{1}{5}} = \frac{1}{5}\sqrt{5}$.

(*c*) *Matrix Space* $\mathbf{M} = \mathbf{M}_{m,n}$: Let $\mathbf{M} = \mathbf{M}_{m,n}$, the vector space of all real $m \times n$ matrices. An inner product is defined on $\mathbf{M}$ by

$$<A, B> = \text{tr}(B^T A)$$

where, as usual, tr() is the trace, i.e., the sum of the diagonal elements. If $A = [a_{ij}]$ and $B = [b_{ij}]$, then

$$\langle A, B \rangle = tr(B^T A) = \sum_{i=1}^{m}\sum_{j=1}^{n} a_{ij}b_{ij} \text{ and } \|A\|^2 = \langle A, A \rangle = \sum_{i=1}^{m}\sum_{j=1}^{n} a_{ij}^2$$

That is, $<A, B>$ is the sum of the corresponding entries in A and B and, in particular, $<A, A>$ is the sum of the squares of the entries of A.

Cauchy-Schwarz Inequality

The following formula is called the Cauchy-Schwarz inequality or Schwarz inequality. It is used in many branches of mathematics.

Theorem 5.1: (**Cauchy-Schwarz**) For any vectors u and v in an inner product space V,

$$\langle u, v \rangle^2 \le \langle u, u \rangle \langle v, v \rangle \text{ or } |\langle u, v \rangle| \le \|u\| \, \|v\|$$

Example 5.3.

(*a*) Consider any real numbers $a_1, \ldots, a_n, b_1, \ldots, b_n$. Then, by the Cauchy-Schwarz inequality,

$$\left(a_1b_1 + a_2b_2 + \ldots + a_nb_n\right)^2 \le \left(a_1^2 + a_2^2 + \ldots + a_n^2\right)\left(b_1^2 + b_2^2 + \ldots + b_n^2\right)$$

That is, $(u \cdot v)^2 \le \|u\|^2\|v\|^2$, where $u = (a_i)$ and $v = (b_i)$.

(*b*) Let *f* and *g* be polynomial functions on the unit interval [0,1]. Then, by the Cauchy-Schwarz inequality,

$$\left[\int_0^1 f(t)g(t)dt\right]^2 \le \int_0^1 f^2(t)dt \int_0^1 g^2(t)dt$$

That is, $(\langle f, g\rangle)^2 \le \|f\|^2\|g\|^2$. Here *V* is the inner product space on **P**(*t*).

Theorem 5.2: Let *V* be an inner product space. Then the norm in *V* satisfies the following properties:

[N_1] $\|v\| \ge 0$; and $\|v\| = 0$ if and only if $v = 0$.
[N_2] $\|kv\| = |k|\|v\|$.
[N_3] $\|u + v\| \le \|u\| + \|v\|$.

The property [N_3] is called the *triangle inequality*; because if we view $u + v$ as the side of the triangle formed with sides *u* and *v*, then [N_3] states that the length of one side of a triangle cannot be greater than the sum of the lengths of the other two sides.

For any nonzero vectors *u* and *v* in an inner product space *V*, the angle between *u* and *v* is defined to be the angle θ such that $0 \le \theta \le \pi$ and $\cos\theta = \dfrac{\langle u, v\rangle}{\|u\|\,\|v\|}$. By the Cauchy-Schwartz inequality, $-1 \le \cos\theta \le 1$, and so the angle exists and is unique.

Example 5.4. Consider the vectors $u = (2, 3, 5)$ and $v = (1, -4, 3)$ in $\mathbf{R}^3$. Then

$$\langle u, v\rangle = 2 - 12 + 15 = 5,$$

$$\|u\| = \sqrt{4 + 9 + 25} = \sqrt{38}, \quad \|v\| = \sqrt{1 + 16 + 9} = \sqrt{26}$$

Then the angle θ between *u* and *v* is given by $\cos\theta = \dfrac{5}{\sqrt{38}\sqrt{26}}$.

Note that θ is an acute angle, since $\cos\theta$ is positive.

Orthogonality

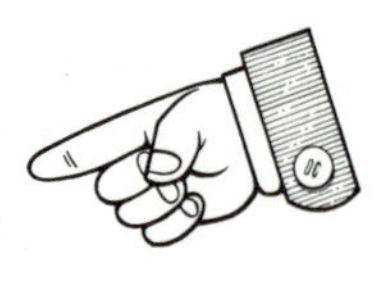

Let V be an inner product space. The vectors $u, v \in V$ are said to be *orthogonal* and u is said to be *orthogonal* to v if $\langle u, v \rangle = 0$. The relation is clearly symmetric, that is, if u is orthogonal to v, then $\langle v,u \rangle = 0$, and so v is orthogonal to u. We note that $0 \in V$ is orthogonal to every $v \in V$, since $\langle 0, v \rangle = \langle 0v, v \rangle = 0\langle v, v \rangle = 0$. Conversely, if u is orthogonal to every $v \in V$, then $\langle u, u \rangle = 0$ and hence $u = 0$ by $[I_3]$. Observe that u and v are orthogonal if and only if $\cos\theta = 0$, where θ is the angle between u and v. Also, this is true if and only if u and v are "perpendicular," i.e., $\theta = \pi/2$ (or $\theta = 90°$).

Example 5.5. Consider the vectors $u = (1, 1, 1)$, $v = (1, 2, -3)$, $w = (1, -4, 3)$ in $\mathbf{R}^3$. Then

$$\langle u, v \rangle = 1 + 2 - 3 = 0 \quad \langle u, w \rangle = 1 - 4 + 3 = 0$$
$$\langle v, w \rangle = 1 - 8 - 9 = -16$$

A vector $w = (x_1, x_2, \ldots, x_n)$ is orthogonal to $u = (a_1, a_2, \ldots, a_n)$ in $\mathbf{R}^n$ if $\langle u, w \rangle = a_1x_1 + a_2x_2 + \ldots + a_nx_n = 0$. That is, w is orthogonal to u if w satisfies a homogeneous equation whose coefficients are the elements of u.

Example 5.6. Find a nonzero vector w that is orthogonal to both

$$u_1 = (1, 2, 1) \text{ and } u_2 = (2, 5, 4) \text{ in } \mathbf{R}^3.$$

Let $w = (x, y, z)$. Then we want $\langle u_1, w \rangle = 0$ and $\langle u_2, w \rangle = 0$. This yields the homogeneous system

$$\begin{aligned} x + 2y + z &= 0 \\ 2x + 5y + 4z &= 0 \end{aligned} \quad \text{or} \quad \begin{aligned} x + 2y + z &= 0 \\ y + 2z &= 0 \end{aligned}$$

Here z is the only free variable in the echelon system. Set $z = 1$ to obtain $y = -2$ and $x = 3$. Thus, $w = (3, -2, 1)$ is a desired nonzero vector orthogonal to u_1 and u_2. Any multiple of w will also be orthogonal to u_1 and u_2. Normalizing w, we obtain the following unit vector orthogonal to u_1 and u_2:

$$\hat{w} = \frac{w}{\|w\|} = \left(\frac{3}{\sqrt{14}}, -\frac{2}{\sqrt{14}}, \frac{1}{\sqrt{14}}\right)$$

Let S be a subset of an inner product space V. The *orthogonal complement* of S, denoted by $S^{\perp}$ (read "S perp") consists of those vectors in V that are orthogonal to every vector $u \in S$; that is,

$$S^{\perp} = \{v \in V: \langle v, u\rangle = 0 \text{ for every } u \in S\}$$

In particular, for a given vector u in V, we have

$$u^{\perp} = \{v \in V: \langle v, u\rangle = 0\}$$

that is, $u^{\perp}$ consists of all vectors in V that are orthogonal to the given vector u.

We show that $S^{\perp}$ is a subspace of V. Clearly $0 \in S^{\perp}$, since 0 is orthogonal to every vector in V. Now suppose $v, w \in S^{\perp}$. Then, for any scalars a and b and any vector $u \in S$, we have

$$\langle av + bw, u\rangle = a\langle v, u\rangle + b\langle w, u\rangle = a \cdot 0 + b \cdot 0 = 0.$$

Thus $av + bw \in S^{\perp}$, and therefore $S^{\perp}$ is a subspace of V.

We state this result formally.

Proposition 5.3: Let S be a subset of a vector space V. Then $S^{\perp}$ is a subspace of V.

Example 5.7.

Find a basis for the subspace $u^{\perp}$ of $\mathbf{R}^3$, where $u = (1, 3, -4)$.

Note that $u^{\perp}$ consists of all vectors $w = (x, y, z)$ such that $\langle u, w\rangle = 0$, or $x + 3y - 4z = 0$. The free variables are y and z.

(1) Set $y = 1$, $z = 0$ to obtain the solution $w_1 = (-3, 1, 0)$.

(2) Set $y = 0$, $z = 1$ to obtain the solution $w_2 = (4, 0, 1)$.

The vectors w_1 and w_2 form a basis for the solution space of the equation, and hence a basis for $u^{\perp}$.

Suppose W is a subspace of V. Then both W and $W^{\perp}$ are subspaces of V. The next theorem is a basic result in linear algebra.

Theorem 5.4: Let W be a subspace of V. Then V is the direct sum of W and $W^{\perp}$, that is, $V = W \oplus W^{\perp}$.

Consider a set $S = \{u_1, u_2, \ldots, u_r\}$ of nonzero vectors in an inner product space V. S is called *orthogonal* if each pair of vectors in S are orthogonal, and S is called *orthonormal* if S is orthogonal and each vector in S has unit length. That is:

(i) Orthogonal: $<u_i, u_j> = 0$ for $i \neq j$

(ii) Orthonormal: $< u_i, u_j > = \begin{cases} 0 & for \quad i \neq j \\ 1 & for \quad i = j \end{cases}$

You Need to Know 

Normalizing an orthogonal set S refers to the process of multiplying each vector in S by the reciprocal of its length in order to transform S into an orthonormal set of vectors.

Theorem 5.5: Suppose S in an orthogonal set of nonzero vectors. Then S is linearly independent.

Theorem 5.6: **(Pythagoras)** Suppose $\{u_1, u_2, \ldots, u_r\}$ is an orthogonal set of vectors. Then $\|u_1 + u_2 + \ldots + u_r\| = \|u_1\| + \|u_2\| + \ldots + \|u_r\|$

Example 5.8. Let $E = \{e_1, e_2, e_3\} = \{(1, 0, 0), (0, 1, 0), (0, 0, 1)\}$ be the usual basis of Euclidean space $\mathbf{R}^3$. It is clear that $<e_1, e_2> = <e_1, e_3> = <e_2, e_3> = 0$ and $<e_1, e_1> = <e_2, e_2> = <e_3, e_3> = 1$. Namely, E is an orthonormal basis of $\mathbf{R}^3$. More generally, the usual basis of $\mathbf{R}^n$ is orthonormal for every n.

Let S consist of the following three vectors in $\mathbf{R}^3$:

$$u_1 = (1, 2, 1), u_2 = (2, 1, -4), u_3 = (3, -2, 1)$$

The reader can verify that the vectors are orthogonal; hence they are linearly independent. Thus S is an orthogonal basis of $\mathbf{R}^3$.

Suppose we want to write $v = (7, 1, 9)$ as a linear combination of u_1,

u_2, u_3. First we set v as a linear combination of u_1, u_2, u_3 using unknowns x_1, x_2, x_3 as follows:

$$v = x_1u_1 + x_2u_2 + x_3u_3$$

or

$$(7, 1, 9) = x_1(1, 2, 1) + x_2(2, 1, -4) + x_3(3, -2, 1) \qquad (*)$$

This method uses the fact that the basis vectors are orthogonal, and the arithmetic is much simpler than the method found in Chapter 3. If we take the inner product of each side of (*) with respect to u_i, we get

$$\langle v, u_i \rangle = \langle x_1u_1 + x_2u_2 + x_3u_3, u_i \rangle = x_i \langle u_i, u_i \rangle$$

or

$$x_i = \frac{\langle v, u_i \rangle}{\langle u_i, u_i \rangle}$$

Here two terms drop out, since u_1, u_2, u_3 are orthogonal. Accordingly,

$$x_1 = \frac{\langle v, u_1 \rangle}{\langle u_1, u_1 \rangle} = \frac{7+2+9}{1+4+1} = \frac{18}{6} = 3$$

$$x_2 = \frac{\langle v, u_2 \rangle}{\langle u_2, u_2 \rangle} = \frac{14+1-36}{4+1+16} = -\frac{21}{21} = -1$$

$$x_3 = \frac{\langle v, u_3 \rangle}{\langle u_3, u_3 \rangle} = \frac{21-2+9}{9+4+1} = \frac{28}{14} = 2$$

Thus, again, we get $v = 3u_1 - u_2 + 2u_3$.

Gram-Schmidt Orthogonalization Process

Suppose $\{v_1, v_2, \ldots, v_n\}$ is a basis of an inner product space V. One can use this basis to construct an orthogonal basis $\{w_1, w_2, \ldots, w_n\}$ of V as follows. Set

$$w_1 = v_1$$

$$w_2 = v_2 - \frac{\langle v_2, w_1 \rangle}{\langle w_1, w_1 \rangle} w_1$$

$$w_3 = v_3 - \frac{\langle v_3, w_1 \rangle}{\langle w_1, w_1 \rangle} w_1 - \frac{\langle v_3, w_2 \rangle}{\langle w_2, w_2 \rangle} w_2$$

....

$$w_n = v_n - \frac{\langle v_n, w_1 \rangle}{\langle w_1, w_1 \rangle} w_1 - \frac{\langle v_n, w_2 \rangle}{\langle w_2, w_2 \rangle} w_2 - \ldots - \frac{\langle v_n, w_{n-1} \rangle}{\langle w_2, w_{n-1} \rangle} w_{n-1}$$

The above construction is known as the *Gram-Schmidt orthogonalization process.*

Each vector w_k is a linear combination of v_k and the preceding w's. Hence one can easily show, by induction, that each w_k is a linear combination of $v_1, v_2, \ldots, v_n$.

Suppose $u_1, u_2, \ldots, u_r$ are linearly independent, and so they form a basis for $U = \text{span}(u_i)$. Applying the Gram-Schmidt orthogonalization process to the u's yields an orthogonal basis for U.

Theorem 5.7: Let $\{v_1, v_2, \ldots, v_n\}$ be any basis of an inner product space V. Then there exists an orthonormal basis $\{u_1, u_2, \ldots, u_n\}$ of V such that the change-of-basis matrix from $\{v_i\}$ to $\{u_i\}$ is triangular, that is, for $k = 1, \ldots, n$, $u_k = a_{k1}v_1 + a_{k2}v_2 + \ldots + a_{kk}v_k$.

Theorem 5.8: Suppose $S = \{w_1, w_2, \ldots, w_r\}$ is an orthogonal basis for a subspace W of a vector space V. Then one may extend S to an orthogonal basis for V, that is, one may find vectors $w_{r+1}, \ldots, w_n$ such that $\{w_1, w_2, \ldots, w_n\}$ is an orthogonal basis for V.

Example 5.9. Apply the Gram-Schmidt orthogonalization process to find an orthogonal basis and then an orthonormal basis for the subspace U of $\mathbf{R}^4$ spanned by $v_1 = (1, 1, 1, 1)$, $v_2 = (1, 2, 4, 5)$, $v_3 = (1, -3, -4, -2)$

(1) First set $w_1 = v_1 = (1, 1, 1, 1)$.

(2) Compute. $w_2 = v_2 - \frac{\langle v_2, w_1 \rangle}{\langle w_1, w_1 \rangle} w_1 = v_2 - \frac{12}{4} w_1 = (-2, -1, 1, 2)$

Set $w_2 = (-2, -1, 1, 2)$.

(3) Compute

$$w_3 = v_3 - \frac{\langle v_3, w_1 \rangle}{\langle w_1, w_1 \rangle} w_1 - \frac{\langle v_3, w_2 \rangle}{\langle w_2, w_2 \rangle} w_2$$

$$= v_3 - \frac{(-8)}{4} w_1 - \frac{(-7)}{10} w_2 = \left(\frac{8}{5}, -\frac{17}{10}, -\frac{13}{10}, \frac{7}{5} \right).$$

Clear fractions to obtain $w_3 = (16, -17, -13, 14)$.
Thus w_1, w_2, w_3 form an orthogonal basis for U. Normalize these vectors to obtain an orthonormal basis $\{u_1, u_2, u_3\}$ of U.

Chapter 6
DETERMINANTS

IN THIS CHAPTER:

- ✔ *Determinants of Order 1, 2, and 3*
- ✔ *Permutations*
- ✔ *Determinants of Arbitrary Order*
- ✔ *Minors and Cofactors*
- ✔ *Evaluation of Determinants*
- ✔ *Cramer's Rule*

Determinants of Order 1, 2, and 3

Each n-square matrix $A = [a_{ij}]$ is assigned a special scalar called the *determinant* of A, denoted by $\det(A)$ or $|A|$ or

$$\begin{vmatrix} a_{11} & a_{12} & \cdots & a_{1n} \\ a_{21} & a_{22} & \cdots & a_{2n} \\ \cdots & \cdots & \cdots & \cdots \\ a_{n1} & a_{n2} & \cdots & a_{nn} \end{vmatrix}$$

We emphasize that an $n \times n$ array of scalars enclosed by straight lines, called a *determinant of order n*, is not a matrix but denotes the determinant of the enclosed array of scalars, i.e., the enclosed matrix.

The determinant function was first discovered during the investigation of systems of linear equations. We shall see that the determinant is an indispensable tool in investigating and obtaining properties of square matrices.

The definition of the determinant and most of its properties also apply in the case where the entries of a matrix come from a commutative ring.

We begin with a special case of determinants of order 1, 2, and 3. Then we define a determinant of arbitrary order. This general definition is preceded by a discussion of permutations, which is necessary for our general definition of the determinant.

Determinants of orders 1 and 2 are defined as follows:

$$|a_{11}| = a_{11} \text{ and } \begin{vmatrix} a_{11} & a_{12} \\ a_{21} & a_{22} \end{vmatrix} = a_{11}a_{22} - a_{12}a_{21}.$$

Thus the determinant of a 1×1 matrix $A = [a_{11}]$ is the scalar a_{11} itself; that is, $\det(A) = |a_{11}| = a_{11}$. The determinant of order two may be remembered by using the following diagram:

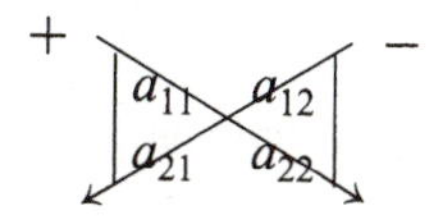

That is, the determinant is equal to the product of the elements along the plus-labeled arrow (the "forward" diagonal) minus the product of the elements along the minus-labeled arrow. (There is an analogous diagram for determinants of order 3, but not for higher-order determinants.)

Example 6.1.
Since the determinant of order one is the scalar itself, we have:

$$\det(27) = 27, \det(-7) = -7, \det(t-3) = t-3$$

$$\begin{vmatrix} 5 & 3 \\ 4 & 6 \end{vmatrix} = 5(6) - 3(4) = 30 - 12 = 18, \quad \begin{vmatrix} 3 & 2 \\ -5 & 7 \end{vmatrix} = 21 + 10 = 31$$

Consider two linear equations in two unknowns, say

$$a_1x + b_1y = c_1$$
$$a_2x + b_2y = c_2$$

Let $D = a_1b_2 - a_2b_1$, the determinant of the matrix of coefficients. Then the system has a unique solution if and only if $D \neq 0$. In such a case, the unique solution may be expressed completely in terms of determinants as follows:

$$x = \frac{N_x}{D} = \frac{b_2c_1 - b_1c_2}{a_1b_2 - a_2b_1} = \frac{\begin{vmatrix} c_1 & b_1 \\ c_2 & b_2 \end{vmatrix}}{\begin{vmatrix} a_1 & b_1 \\ a_2 & b_2 \end{vmatrix}},$$

$$y = \frac{N_y}{D} = \frac{a_2c_1 - a_1c_2}{a_1b_2 - a_2b_1} = \frac{\begin{vmatrix} a_1 & c_1 \\ a_2 & c_2 \end{vmatrix}}{\begin{vmatrix} a_1 & b_1 \\ a_2 & b_2 \end{vmatrix}}$$

Here D appears in the denominator of both quotients. The numerators N_x and N_y of the quotients for x and y, respectively, can be obtained by substituting the column of constant terms in place of the column of coefficients of the given unknown in the matrix of coefficients. On the other hand, if $D = 0$, then the system may have no solution or more than one solution.

Example 6.2. Solve by determinants the system $\begin{cases} 4x - 3y = 15 \\ 2x + 5y = 1 \end{cases}$

First find the determinant D of the matrix of coefficients:

$$D = \begin{vmatrix} 4 & -3 \\ 2 & 5 \end{vmatrix} = 4(5) - (-3)(2) = 20 + 6 = 26$$

Since $D \neq 0$, the system has a unique solution. To obtain the numerators N_x and N_y simply replace, in the matrix of coefficients, the coefficients of x and y, respectively, by the constant terms, and then take their determinants:

$$N_x = \begin{vmatrix} 15 & -3 \\ 1 & 5 \end{vmatrix} = 75 + 3 = 78, \quad N_y = \begin{vmatrix} 4 & 15 \\ 2 & 1 \end{vmatrix} = 4 - 30 = -26$$

Then the unique solution of the system is

$$x = \frac{N_x}{D} = \frac{78}{26} = 3 \quad y = \frac{N_y}{D} = -\frac{26}{26} = -1$$

Consider an arbitrary 3×3 matrix $A = [a_{ij}]$. The determinant of A is defined as follows:

$$\det(A) = \begin{vmatrix} a_{11} & a_{12} & a_{13} \\ a_{21} & a_{22} & a_{23} \\ a_{31} & a_{32} & a_{33} \end{vmatrix} = \begin{matrix} (a_{11}a_{22}a_{33} + a_{12}a_{23}a_{31} + a_{13}a_{21}a_{32}) \\ -(a_{13}a_{22}a_{31} + a_{12}a_{21}a_{33} + a_{11}a_{23}a_{32}) \end{matrix}$$

Observe that there are six products, each product consisting of three elements of the original matrix. Three of the products are added and three of the products are subtracted.

The diagrams in Figure 6-1 may help to remember the above six products in det(A). That is, the determinant is equal to the sum of the products of the elements along the three plus-labeled arrows in Figure 6-1 plus the sum of the negatives of the products of the elements along the three minus-labeled arrows. We emphasize that there are no such diagrammatic devices to remember determinants of higher order.

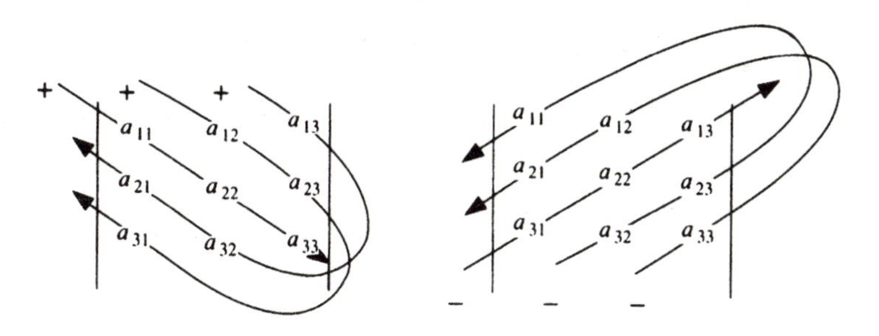

Figure 6.1

Example 6.3. Let $A = \begin{bmatrix} 2 & 1 & 1 \\ 0 & 5 & -2 \\ 1 & -3 & 4 \end{bmatrix}$ and $B = \begin{bmatrix} 3 & 2 & 1 \\ -4 & 5 & -1 \\ 2 & -3 & 4 \end{bmatrix}$. Find det($A$) and det($B$).

$$\begin{aligned} \det(A) &= 2(5)(4) + 1(-2)(1) + 1(0)(-3) - 1(5)(1) - (-2)(-3)(2) - 1(0)(4) \\ &= 40 - 2 + 0 - 5 - 12 - 0 = 21 \\ \det(B) &= 3(5)(4) + 2(-1)(2) + 1(-4)(-3) - 1(5)(2) - 3(-1)(-3) - 2(-4)(4) \\ &= 60 - 4 + 12 - 10 - 9 + 32 = 81 \end{aligned}$$

The determinant of the 3×3 matrix $A = [a_{ij}]$ may be rewritten as follows:

$$A = a_{11}\begin{vmatrix} a_{22} & a_{23} \\ a_{32} & a_{33} \end{vmatrix} - a_{12}\begin{vmatrix} a_{21} & a_{23} \\ a_{31} & a_{33} \end{vmatrix} + a_{13}\begin{vmatrix} a_{21} & a_{22} \\ a_{31} & a_{32} \end{vmatrix}$$
$$= a_{11}(a_{22}a_{33} - a_{23}a_{32}) - a_{12}(a_{21}a_{33} - a_{23}a_{31}) + a_{13}(a_{21}a_{32} - a_{22}a_{31})$$

which is a linear combination of three determinants of order 2 whose coefficients (with alternating signs) form the first row of the given matrix. This linear combination may be indicated in the form

$$a_{11}\left|\begin{array}{c:cc} a_{11} & a_{12} & a_{13} \\ \hdashline a_{21} & a_{22} & a_{23} \\ a_{31} & a_{32} & a_{33} \end{array}\right| - a_{12}\left|\begin{array}{c:c:c} a_{11} & a_{12} & a_{13} \\ \hdashline a_{21} & a_{22} & a_{23} \\ a_{31} & a_{32} & a_{33} \end{array}\right| + a_{13}\left|\begin{array}{cc:c} a_{11} & a_{12} & a_{13} \\ \hdashline a_{21} & a_{22} & a_{23} \\ a_{31} & a_{32} & a_{33} \end{array}\right|$$

Note that each 2×2 matrix can be obtained by deleting, in the original matrix, the row and column containing its coefficient.

Example 6.4.

$$\begin{vmatrix} 1 & 2 & 3 \\ 4 & -2 & 3 \\ 0 & 5 & -1 \end{vmatrix} = 1\left|\begin{array}{c:cc} 1 & 2 & 3 \\ \hdashline 4 & -2 & 3 \\ 0 & 5 & -1 \end{array}\right| - 2\left|\begin{array}{c:c:c} 1 & 2 & 3 \\ \hdashline 4 & -2 & 3 \\ 0 & 5 & -1 \end{array}\right| + 3\left|\begin{array}{cc:c} 1 & 2 & 3 \\ \hdashline 4 & -2 & 3 \\ 0 & 5 & -1 \end{array}\right|$$
$$= 1\begin{vmatrix} -2 & 3 \\ 5 & -1 \end{vmatrix} - 2\begin{vmatrix} 4 & 3 \\ 0 & -1 \end{vmatrix} + 3\begin{vmatrix} 4 & -2 \\ 0 & 5 \end{vmatrix}$$
$$= 1(2-15) - 2(-4-0) + 3(20-0) = -13 + 8 + 60 = 55$$

Permutations

A *permutation* σ of the set $(1, 2, \ldots, n)$ is a one-to-one mapping of the set onto itself or, equivalently, a rearrangement of the numbers $1, 2, \ldots, n$. Such a *permutation* σ is denoted by

$$\sigma = \begin{pmatrix} 1 & 2 & \ldots & n \\ j_1 & j_2 & \cdots & j_n \end{pmatrix} \quad or \quad \sigma = j_1 j_2 \cdots j_n \text{ where } j_i = \sigma(i)$$

The set of all such permutations is denoted by S_n, and the number of such permutations is $n!$. If $\sigma \in S_n$, then the inverse mapping $\sigma^{-1} \in S_n$; and if σ, $\tau \in S_n$, then the composition mapping $\sigma \circ \tau \in S_n$. Also, the identity mapping $\varepsilon = \sigma \circ \sigma^{-1} \in S_n$. (In fact, $\varepsilon = 123\ldots n$.)

Example 6.5.

(*a*) There are $2! = 2 \cdot 1 = 2$ permutations in S_2; they are 12 and 21.
(*b*) There are $3! = 3 \cdot 2 \cdot 1 = 6$ permutations in S_3; they are 123, 132, 213, 231, 312, 321.

Consider an arbitrary permutation σ in S_n, say $\sigma = j_1 j_2 \ldots j_n$. We say σ is an even or odd permutation according to whether there is an even or odd number of inversions in σ. By an inversion in σ we mean a pair of integers (i, k) such that $i > k$, but i precedes k in σ. We then define the sign or parity of σ, written sgn σ, by

$$\operatorname{sgn} \sigma = \begin{cases} 1 \text{ if } \sigma \text{ is even} \\ -1 \text{ if } \sigma \text{ is odd} \end{cases}$$

Example 6.6.

(*a*) Find the sign of $\sigma = 35142$ in S_5.
For each element k, we count the number of elements i such that $i > k$ and i precedes k in σ. There are:
2 numbers (3 and 5) greater than and preceding 1,
3 numbers (3, 5, and 4) greater than and preceding 2,
1 number (5) greater than and preceding 4.
(There are no numbers that are greater than and preceding either 3 or 5.) Since there are, in all, six inversions, σ is even and sgn $\sigma = 1$.

(*b*) Let τ be the permutation that interchanges two numbers i and j and leaves the other numbers fixed. That is,

$$\tau(i) = j,\ \tau(j) = i,\ \tau(k) = k \text{ where } k \neq i, j.$$

We call τ a *transposition*. If $i < j$, then there are $2(j - i) + 1$ inversions in τ, and hence the transposition τ is odd.

Determinants of Arbitrary Order

Let $A = [a_{ij}]$ be a square matrix of order n over a field K.

Consider a product of n elements of A such that one and only one element comes from each row and one and only one element comes from each column. Such a product can be written in the form

$$a_{1j_1}, a_{2j_2}, \ldots, a_{nj_n}$$

that is, where the factors come from successive rows, and so the first subscripts are in the natural order 1, 2, … , n. Now since the factors come from different columns, the sequence of second subscripts forms a permutation $\sigma = j_1 j_2 \ldots j_n$ in S_n. Conversely, each permutation in S_n determines a product of the above form. Thus the matrix A contains $n!$ such products.

The *determinant* of $A = [a_{ij}]$, denoted by det(A) or $|A|$, is the sum of all the above $n!$ products, where each such product is multiplied by sgn σ. That is,

$$|A| = \sum_{\sigma} (\operatorname{sgn} \sigma) a_{1j_1} a_{2j_2} \ldots a_{nj_n}$$

or

$$|A| = \sum_{\sigma \in S_n} (\operatorname{sgn} \sigma) a_{1\sigma(1)} a_{2\sigma(2)} \ldots a_{n\sigma(n)}$$

The determinant of the n-square matrix A is said to be of order n.

The next example shows that the above definition agrees with the previous definition of determinants of order 1, 2, and 3.

Example 6.7.

(*a*) Let $A = [a_{11}]$ be a 1×1 matrix. Since S_1 has only one permutation, which is even, det(A) = a_{11}, the number itself.

(*b*) Let $A = [a_{ij}]$ be a 2×2 matrix. In S_2, the permutation 12 is even and the permutation 21 is odd. Hence

$$\det(A) = \begin{vmatrix} a_{11} & a_{12} \\ a_{21} & a_{22} \end{vmatrix} = a_{11}a_{22} - a_{12}a_{21}$$

(*c*) Let $A = [a_{ij}]$ be a 3×3 matrix. In S_3, the permutations 123, 231, 312 are even and the permutations 132, 213, 321 are odd. Hence

$$\det(A) = \begin{vmatrix} a_{11} & a_{12} & a_{13} \\ a_{21} & a_{22} & a_{23} \\ a_{31} & a_{32} & a_{33} \end{vmatrix}$$

$$= a_{11}a_{22}a_{33} + a_{12}a_{23}a_{31} + a_{13}a_{21}a_{32} - a_{11}a_{23}a_{32} - a_{12}a_{21}a_{33} - a_{13}a_{22}a_{31}$$

As n increases, the number of terms in the determinant becomes astronomical. Accordingly, we use indirect methods to evaluate determinants rather than the definition of the determinant. In fact, we prove a number of properties about determinants that will permit us to shorten the computation considerably. In particular, we show that a determinant of order n is equal to a linear combination of determinants of order $n - 1$, as in the case $n = 3$ above.

We now list basic properties of the determinant.

Theorem 6.1: The determinant of a matrix A and its transpose A^T are equal; that is, $|A| = |A^T|$.

By this theorem any theorem about the determinant of a matrix A that concerns the rows of A will have an analogous theorem concerning the columns of A.

The next theorem gives certain cases for which the determinant can be obtained immediately:

Theorem 6.2: Let A be a square matrix.

(i) If A has a row (column) of zeros, then $|A| = 0$.
(ii) If A has two identical rows (columns), then $|A| = 0$.
(iii) If A is triangular, i.e., A has zeros above or below the diagonal, then $|A|$ = product of diagonal elements. Thus in particular, $|I| = 1$, where I is the identity matrix.

The next theorem shows how the determinant of a matrix is affected by the elementary row and column operations.

Theorem 6.3: Suppose B is obtained from A by an elementary row (column) operation.

(i) If two rows (columns) of A were interchanged, then $|B| = -|A|$.
(ii) If a row (column) of A were multiplied by a scalar k, then $|B| = k|A|$.
(iii) If a multiple of a row (column) of A were added to another row (column) of A, then $|B| = |A|$.

We now state two of the most important and useful theorems on determinants.

Theorem 6.4: The determinant of a product of two matrices A and B is the product of their determinants; that is, $\det(AB) = \det(A)\det(B)$.

Theorem 6.5: Let A be a square matrix. Then the following are equivalent:

(i) A is invertible; that is, A has an inverse A^{-1}.
(ii) $AX = 0$ has only the zero solution.
(iii) The determinant of A is not zero; that is, $\det(A) \neq 0$.

Depending on the author and the text, a nonsingular matrix A is defined to be an invertible matrix A, or a matrix A for which $|A| \neq 0$, or a matrix A for which $AX = 0$ has only the zero solution. The above theorem shows that all such definitions are equivalent.

Matrices A and B are *similar* if there exists a nonsingular matrix P such that $B = P^{-1}AP$. Using the multiplicative property of the determinant (Theorem 6.4), one can easily prove the following theorem.

Theorem 6.6: Suppose A and B are similar matrices. Then $|A| = |B|$.

Minors and Cofactors

Consider an n-square matrix $A = [a_{ij}]$. Let M_{ij} denote the $(n - 1)$-square submatrix of A obtained by deleting its ith row and jth column. The determinant $|M_{ij}|$ is called the *minor* of the element a_{ij} of A, and we define the *cofactor* of a_{ij}, denoted by A_{ij}, to be the "signed" minor:

$$A_{ij} = (-1)^{i+j}|M_{ij}|$$

Note that the "signs" $(-1)^{i+j}$ accompanying the minors form a chessboard pattern with +'s on the main diagonal:

$$\begin{bmatrix} + & - & + & - & \cdots \\ - & + & - & + & \cdots \\ + & - & + & - & \cdots \\ \cdots & \cdots & \cdots & \cdots & \cdots \end{bmatrix}$$

We emphasize that M_{ij} denotes a matrix whereas A_{ij} denotes a scalar.

Example 6.8. Let $A = \begin{bmatrix} 1 & 2 & 3 \\ 4 & 5 & 6 \\ 7 & 8 & 9 \end{bmatrix}$. Find the following minors and cofactors: (*a*) $|M_{23}|$ and A_{23}, (*b*) $|M_{31}|$ and A_{31}.

(*a*) $|M_{23}| = \begin{vmatrix} 1 & 2 & 3 \\ 4 & 5 & 6 \\ 7 & 8 & 9 \end{vmatrix} = \begin{vmatrix} 1 & 2 \\ 7 & 8 \end{vmatrix} = 8 - 14 = -6,$

and so $A_{23} = (-1)^{2+3}|M_{23}| = -(-6) = 6.$

(*b*) $|M_{31}| = \begin{vmatrix} 1 & 2 & 3 \\ 4 & 5 & 6 \\ 7 & 8 & 9 \end{vmatrix} = \begin{vmatrix} 2 & 3 \\ 5 & 6 \end{vmatrix} = 12 - 15 = -3,$

and so $A_{31} = (-1)^{3+1}|M_{31}| = +(-3) = -3.$

Minors and Cofactors

Let $A = [a_{ij}]$ be a square matrix of order n. Consider any r rows and r columns of A. That is, consider any set $I = (i_1, i_2, \ldots, i_r)$ of r row indices and any set $J = (j_1, j_2, \ldots, j_r)$ of r column indices. Then I and J define an $r \times r$ submatrix of A, denoted by $A(I;J)$, obtained by deleting the rows and columns of A whose subscripts do not belong to I or J, respectively. That is, $A(I;J) = [a_{st} : s \in I, t \in J]$. The determinant $|A(I;J)|$ is called a *minor* of A of order r and

$$(-1)^{i_1+i_2+\ldots i_r+j_1+j_2+\ldots j_r}|A(I;J)|$$

is the corresponding signed minor. (Note that a minor of order $n-1$ is a minor of the element a_{ij} and the corresponding signed minor is a cofactor.) Furthermore, if I' and J' denote, respectively, the remaining row and column indices, then $|A(I';J')|$ denotes the *complementary minor*, and its sign is the same sign as the minor itself.

Example 6.9. Let $A = [a_{ij}]$ be a 5-square matrix, and let $I = \{1, 2, 4\}$ and $J = \{2, 3, 5\}$. Then $I' = \{3, 5\}$ and $J' = \{1, 4\}$, and the corresponding minor $|M|$ and complementary minor $|M'|$ are as follows:

$$|M| = |A(I;J)| = \begin{vmatrix} a_{12} & a_{13} & a_{15} \\ a_{22} & a_{23} & a_{25} \\ a_{42} & a_{43} & a_{45} \end{vmatrix} \text{ and } |M'| = |A(I';J')| = \begin{vmatrix} a_{31} & a_{34} \\ a_{51} & a_{54} \end{vmatrix}$$

Since $1 + 2 + 4 + 2 + 3 + 5 = 17$ is odd, $-|M|$ is the signed minor, and $-|M'|$ is the signed complementary minor.

A minor is *principal* if the row and column indices are the same, or equivalently, if the diagonal elements of the minor come from the diagonal of the matrix. We note that the sign of a principal minor is always +1 since the sum of the row and identical column subscripts must always be even.

Example 6.10. Let $A = \begin{bmatrix} 1 & 2 & -1 \\ 3 & 5 & 4 \\ -3 & 1 & -2 \end{bmatrix}$. Find the sums C_1, C_2, and C_3 of the principal minors of A orders one, two and three, respectively.

(*a*) There are three principal minors of order one. These are $|1| = 1$, $|5| = 5$, $|-2| = -2$, and so $C_1 = 1 + 5 - 2 = 4$
Note that C_1 is simply the trace of A. Namely, $C_1 = \text{tr}(A)$.

(*b*) There are three ways to choose two of the three diagonal elements, and each choice gives a minor of order two. These are

$$\begin{vmatrix} 1 & 2 \\ 3 & 5 \end{vmatrix} = 5 - 6 = -1, \quad \begin{vmatrix} 1 & -1 \\ -3 & -2 \end{vmatrix} = -2 - 3 = -5, \quad \begin{vmatrix} 5 & 4 \\ 1 & -2 \end{vmatrix} = -10 - 4 = -14$$

(Note that these minors of order two are the cofactors A_{33}, A_{22}, and A_{11} of A respectively.) Thus $C_2 = -1 - 5 - 14 = -20$.

(c) There is only one way to choose three of the three diagonal elements. Thus the only minor of order three is the determinant of A itself. Thus $C_3 = |A| = -10 - 24 - 3 - 15 + 12 - 4 = -44$.

Theorem 6.7: (**Laplace**) The determinant of a square matrix $A = [a_{ij}]$ is equal to the sum of the products obtained by multiplying the elements of any row (column) by their respective cofactors:

$$|A| = a_{i1}A_{i1} + a_{i2}A_{i2} + \ldots + a_{in}A_{in} = \sum_{j=1}^{n} a_{ij}A_{ij}$$

$$|A| = a_{1j}A_{1j} + a_{2j}A_{2j} + \ldots + a_{nj}A_{nj} = \sum_{i=1}^{n} a_{ij}A_{ij}$$

The above formulas for $|A|$ are called the *Laplace expansions* of the determinant of A by the ith row and the jth column. Together with the elementary row (column) operations, they offer a method of simplifying the computation of $|A|$, as described below.

Evaluation of Determinants

The following algorithm reduces the evaluation of a determinant of order n to the evaluation of a determinant of order $n - 1$.

Algorithm 6.1: (**Reduction of the order of a determinant**) The input is a nonzero n-square matrix $A = [a_{ij}]$ with $n > 1$.

Step 1. Choose an element $a_{ij} = 1$ or, if lacking, $a_{ij} \neq 0$.
Step 2. Using a_{ij} as a pivot, apply elementary row (column) operations to put 0's in all the other positions in the column (row) containing a_{ij}.
Step 3. Expand the determinant by the column (row) containing a_{ij}.

Algorithm 6.1 is usually used for determinants of order 4 or more. With determinants of order less than 4, one uses the specific formulas for the determinant.

Gaussian elimination or, equivalently, repeated use of Algorithm 6.1 together with row interchanges can be used to transform a matrix A into an upper triangular matrix whose determinant is the product of its diagonal entries. However, one must keep track of the number of row interchanges, since each row interchange changes the sign of the determinant.

Example 6.11. Use Algorithm 6.1 to find the determinant of

$$A = \begin{bmatrix} 5 & 4 & 2 & 1 \\ 2 & 3 & 1 & -2 \\ -5 & -7 & -3 & 9 \\ 1 & -2 & -1 & 4 \end{bmatrix}$$

Use $a_{23} = 1$ as a pivot to put 0's in the other positions of the third column, that is, apply the row operations "Replace R_1 by $-2R_2 + R_1$," "Replace R_3 by $3R_2 + R_3$," and "Replace R_4 by $R_2 + R_4$." By Theorem 6.3 (iii), the value of the determinant does not change under these operations. Thus

$$|A| = \begin{vmatrix} 5 & 4 & 2 & 1 \\ 2 & 3 & 1 & -2 \\ -5 & -7 & -3 & 9 \\ 1 & -2 & -1 & 4 \end{vmatrix} = \begin{vmatrix} 1 & -2 & 0 & 5 \\ 2 & 3 & 1 & -2 \\ 1 & 2 & 0 & 3 \\ 3 & 1 & 0 & 2 \end{vmatrix}$$

Now expand by the third column. Specifically, neglect all terms that contain 0 and use the fact that the sign of the minor M_{23} is $(-1)^{2+3} = -1$. Thus

$$|A| = \left|\begin{array}{cc:c:c} 1 & -2 & 0 & 5 \\ \hdashline 2 & 3 & 1 & -2 \\ \hdashline 1 & 2 & 0 & 3 \\ 3 & 1 & 0 & 2 \end{array}\right| = -\begin{vmatrix} 1 & -2 & 5 \\ 1 & 2 & 3 \\ 3 & 1 & 2 \end{vmatrix} = -(4 - 18 + 5 - 30 + 4 - 3) = 38$$

Theorem 6.8: Suppose M is an upper (lower) triangular block matrix with the diagonal blocks $A_1, A_2, \ldots, A_n$. Then

$$\det(M) = \det(A_1)\det(A_2)\ldots\det(A_n).$$

Example 6.12. Find $|M|$ where $M = \left[\begin{array}{cc:ccc} 2 & 3 & 4 & 7 & 8 \\ -1 & 5 & 3 & 2 & 1 \\ \hdashline 0 & 0 & 2 & 1 & 5 \\ 0 & 0 & 3 & -1 & 4 \\ 0 & 0 & 5 & 2 & 6 \end{array}\right]$.

Note that M is an upper triangular block matrix. Evaluate the determinant of each diagonal block:

$$\begin{vmatrix} 2 & 3 \\ -1 & 5 \end{vmatrix} = 10 + 3 = 13, \begin{vmatrix} 2 & 1 & 5 \\ 3 & -1 & 4 \\ 5 & 2 & 6 \end{vmatrix} = -12 + 20 + 30 + 25 - 18 - 16 = 29$$

Then $|M| = 13(29) = 377$.

Note!

Suppose $M = \begin{bmatrix} A & B \\ C & D \end{bmatrix}$, where A, B, C, D are square matrices. Then it is *not* generally true that $|M| = |A||D| - |B||C|$.

Cramer's Rule

Consider a system $AX = B$ of n linear equations in n unknowns. Here $A = [a_{ij}]$ is the (square) matrix of coefficients and $B = [b_i]$ is the column vector of constants. Let A_i be the matrix obtained from A by replacing the ith column of A by the column vector B. Furthermore, let

$$D = \det(A),\ N_1 = \det(A_1),\ N_2 = \det(A_2),\ \ldots,\ N_n = \det(A_n)$$

The fundamental relationship between determinants and the solution of the system $AX = B$ follows.

Theorem 6.9: The (square) system $AX = B$ has a solution if and only if $D \neq 0$. In this case, the unique solution is given by

$$x_1 = \frac{N_1}{D}, \quad x_2 = \frac{N_2}{D}, \quad \ldots \quad x_n = \frac{N_n}{D}$$

The above theorem is known as *Cramer's rule* for solving systems of linear equations. We emphasize that the theorem only refers to a system with the same number of equations as unknowns, and that it only gives the solution when $D \neq 0$. In fact, if $D = 0$, the theorem does not tell us whether or not the system has a solution. However, in the case of a homogeneous system, we have the following useful result.

Theorem 6.10: A square homogeneous system $AX = 0$ has a nonzero solution if and only if $D = |A| = 0$.

Example 6.13. Solve using determinants the system $\begin{cases} x + y + z = 5 \\ x - 2y - 3z = -1 \\ 2x + y - z = 3 \end{cases}$

First compute the determinant D of the matrix of coefficients:

$$D = \begin{vmatrix} 1 & 1 & 1 \\ 1 & -2 & -3 \\ 2 & 1 & -1 \end{vmatrix} = 2 - 6 + 1 + 4 + 1 + 3 + 1 = 5$$

Since $D \neq 0$, the system has a unique solution. To compute N_x, N_y, N_z, we replace, respectively, the coefficients of x, y, z in the matrix of coefficients by the constant terms. This yields

$$N_x = \begin{vmatrix} 5 & 1 & 1 \\ -1 & -2 & -3 \\ 3 & 1 & -1 \end{vmatrix} = 20, \quad N_y = \begin{vmatrix} 1 & 5 & 1 \\ 1 & -1 & -3 \\ 2 & 3 & -1 \end{vmatrix} = -10, \quad N_z = \begin{vmatrix} 1 & 1 & 5 \\ 1 & -2 & -1 \\ 2 & 1 & 3 \end{vmatrix} = 15$$

Thus the unique solution of the system is $x = N_x/D = 4$, $y = N_y/D = -2$, $z = N_z/D = 3$, that is, the vector $u = (4, -2, 3)$.

Chapter 7

DIAGONALIZATION: EIGENVALUES AND EIGENVECTORS

IN THIS CHAPTER:

- ✔ *Introduction*
- ✔ *Characteristic Polynomial*
- ✔ *Diagonalization, Eigenvalues, and Eigenvectors*
- ✔ *Computing Eigenvalues and Eigenvectors; Diagonalizing Matrices*
- ✔ *Minimal Polynomials*

Introduction

Suppose an n-square matrix A is given. The matrix A is said to be *diagonalizable* if there exists a nonsingular matrix P such that

$$B = P^{-1}AP$$

is diagonal. This chapter discusses the diagonalization of a matrix A. In particular, an algorithm is given to find the matrix P when it exists.

Consider a polynomial $f(t) = a_n t^n + \ldots + a_1 t + a_0$ over a field K. Recall from Chapter 2 that if A is any square matrix, then we define

$$f(A) = a_n A^n + \ldots + a_1 A + a_0 I$$

Where I is the identity matrix. In particular, we say that A is a *root* of $f(A) = 0$, the zero matrix.

Example 7.1. Let $A = \begin{bmatrix} 1 & 2 \\ 3 & 4 \end{bmatrix}$. Then $A^2 = \begin{bmatrix} 7 & 10 \\ 15 & 22 \end{bmatrix}$. If

$$f(t) = 2t^2 - 3t + 5 \text{ and } g(t) = t^2 - 5t - 2$$

then

$$f(A) = 2A^2 - 3A + 5I = \begin{bmatrix} 14 & 20 \\ 30 & 44 \end{bmatrix} + \begin{bmatrix} -3 & -6 \\ -9 & -12 \end{bmatrix} + \begin{bmatrix} 5 & 0 \\ 0 & 5 \end{bmatrix} = \begin{bmatrix} 16 & 14 \\ 21 & 37 \end{bmatrix}$$

and

$$g(A) = A^2 - 5A - 2I = \begin{bmatrix} 7 & 10 \\ 15 & 22 \end{bmatrix} + \begin{bmatrix} -5 & -10 \\ -15 & -20 \end{bmatrix} + \begin{bmatrix} -2 & 0 \\ 0 & -2 \end{bmatrix} = \begin{bmatrix} 0 & 0 \\ 0 & 0 \end{bmatrix}$$

Thus A is a zero of $g(t)$.

Theorem 7.1: Let f and g be polynomials. For any square matrix A and scalar k,

(i) $(f + g)(A) = f(A) + g(A)$ (iii) $(kf)(A) = kf(A)$
(ii) $(fg)(A) = f(A)g(A)$ (iv) $f(A)g(A) = g(A)f(A)$

Observe that (iv) tells us that any two polynomials in A commute.

Characteristic Polynomial

Let $A = [a_{ij}]$ be an n-square matrix. The matrix $M = A - tI_n$, where I_n is the n-square identity matrix and t is an indeterminate, may be obtained by subtracting t down the diagonal of A. The negative of M is the matrix $tI_n - A$, and its determinant

$$\Delta(t) = \det(tI_n - A) = (-1)^n \det(A - tI_n)$$

which is a polynomial in t of degree n, is called the *characteristic polynomial* of A.

Theorem 7.2: (**Cayley-Hamilton**) Every matrix A is a root of its characteristic polynomial.

Suppose $A = [a_{ij}]$ is a triangular matrix. Then $tI - A$ is a triangular matrix with diagonal entries $t - a_{ii}$; and hence

$$\Delta(t) = \det(tI - A) = (t - a_{11})(t - a_{22}) \ldots (t - a_{nn})$$

Observe that the roots of $\Delta(t)$ are the diagonal elements of A.

Example 7.2. Let $A = \begin{bmatrix} 1 & 3 \\ 4 & 5 \end{bmatrix}$. Its characteristic polynomial is

$$\Delta(t) = |tI - A| = \begin{vmatrix} t-1 & -3 \\ -4 & t-5 \end{vmatrix} = (t-1)(t-5) - 12 = t^2 - 6t - 7$$

As expected from the Cayley-Hamilton Theorem, A is a root of $\Delta(t)$; that is,

$$\Delta(A) = A^2 - 6A - 7I = \begin{bmatrix} 13 & 18 \\ 24 & 37 \end{bmatrix} + \begin{bmatrix} -6 & -18 \\ -24 & -30 \end{bmatrix} + \begin{bmatrix} -7 & 0 \\ 0 & -7 \end{bmatrix} = \begin{bmatrix} 0 & 0 \\ 0 & 0 \end{bmatrix}$$

Now suppose A and B are similar matrices, say $B = P^{-1}AP$, where P is invertible. We show that A and B have the same characteristic polynomial. Using $tI = P^{-1}tIP$, we have

$$\Delta_B(t) = \det(tI - B) = \det(tI - P^{-1}AP) = \det(P^{-1}tIP - P^{-1}AP)$$
$$= \det[P^{-1}(tI - A)P] = \det(P^{-1})\det(tI - A)\det(P)$$

Using the fact that determinants are scalars and commute and that $\det(P^{-1})\det(P) = 1$, we finally obtain $\Delta_B(t) = \det(tI - A) = \Delta_A(t)$.

Thus we have proved the following theorem.

Theorem 7.3: Similar matrices have the same characteristic polynomial.

There are simple formulas for the characteristic polynomials of matrices of orders 2 and 3.

(*a*) Suppose $A = \begin{bmatrix} a_{11} & a_{12} \\ a_{21} & a_{22} \end{bmatrix}$. Then

$$\Delta(t) = t^2 - (a_{11} + a_{22})t + \det(A) = t^2 - \operatorname{tr}(A)t + \det(A)$$

(*b*) Suppose $A = \begin{bmatrix} a_{11} & a_{12} & a_{13} \\ a_{21} & a_{22} & a_{23} \\ a_{31} & a_{32} & a_{33} \end{bmatrix}$. Then

$$\Delta(t) = t^3 - \operatorname{tr}(A)t^2 + (A_{11} + A_{22} + A_{33})t - \det(A)$$

(Here A_{11}, A_{22}, A_{33} denote, respectively, the cofactors of a_{11}, a_{22}, a_{33}.)

Example 7.3. Find the characteristic polynomial of each of the following matrices:

$$(a)\quad A = \begin{bmatrix} 5 & 3 \\ 2 & 10 \end{bmatrix}, \quad (b)\quad B = \begin{bmatrix} 7 & -1 \\ 6 & 2 \end{bmatrix}, \quad (c)\quad C = \begin{bmatrix} 5 & -2 \\ 4 & -4 \end{bmatrix}.$$

(*a*) We have $\operatorname{tr}(A) = 5 + 10 = 15$ and $|A| = 50 - 6 = 44$; hence $\Delta(t) = t^2 - 15t + 44$.

(*b*) We have $\operatorname{tr}(B) = 7 + 2 = 9$ and $|B| = 14 + 6 = 20$; hence $\Delta(t) = t^2 - 9t + 20$.

(*c*) We have $\operatorname{tr}(C) = 5 - 4 = 1$ and $|C| = -20 + 8 = -12$; hence $\Delta(t) = t^2 - t - 12$.

Example 7.4. Find the characteristic polynomial of $A = \begin{bmatrix} 1 & 1 & 2 \\ 0 & 3 & 2 \\ 1 & 3 & 9 \end{bmatrix}$.

We have $\operatorname{tr}(A) = 1 + 3 + 9 = 13$. The cofactors of the diagonal elements are as follows:

$$A_{11} = \begin{vmatrix} 3 & 2 \\ 3 & 9 \end{vmatrix} = 21, \quad A_{22} = \begin{vmatrix} 1 & 2 \\ 1 & 9 \end{vmatrix} = 7, \quad A_{33} = \begin{vmatrix} 1 & 1 \\ 0 & 3 \end{vmatrix} = 3.$$

Thus $A_{11} + A_{22} + A_{33} = 31$. Also, $|A| = 27 + 2 + 0 - 6 - 6 - 0 = 17$. Accordingly, $\Delta(t) = t^3 - 13t^2 + 31t - 17$.

Theorem 7.4: Let A be an n-square matrix. Then its characteristic polynomial is $\Delta(t) = t^n - S_1 t^{n-1} + S_2 t^{n-2} + \ldots + (-1)^n S_n$ where S_k is the sum of the principal minors of order k.

Diagonalization, Eigenvalues, and Eigenvectors

Let A be any n-square matrix. Then A can be represented by (or is similar to) a diagonal matrix $D = \text{diag}(k_1, k_2, \ldots, k_n)$ if and only if there exists a basis S consisting of (column) vectors $u_1, u_2, \ldots, u_n$ such that

$$\begin{aligned} Au_1 &= k_1 u_1 \\ Au_2 &= k_2 u_2 \\ &\cdots\cdots \\ Au_n &= k_n u_n \end{aligned}$$

In such a case, A is said to be *diagonalizable*. Furthermore, $D = P^{-1}AP$, where P is the nonsingular matrix whose columns are, respectively, the basis vectors $u_1, u_2, \ldots, u_n$.

Let A be any square matrix. A scalar λ is called an *eigenvalue* of A if there exists a nonzero (column) vector v such that $Av = \lambda v$. Any vector satisfying this relation is called an *eigenvector* of A *belonging* to the eigenvalue λ.

We note that each scalar multiple kv of an eigenvector v belonging to λ is also such an eigenvector, since $A(kv) = k(\lambda v) = \lambda(kv)$. The set E_λ of all such eigenvectors is a subspace of V, called the *eigenspace* of λ. (If $\dim E_\lambda = 1$, then E_λ is called an *eigenline* and λ is called a *scaling factor*.)

You Need to Know 

The terms *characteristic value* and *characteristic vector* (or *proper value* and *proper vector*) are sometimes used instead of eigenvalue and eigenvector.

Theorem 7.5: An n-square matrix A is similar to a diagonal matrix D if and only if A has n linearly independent eigenvectors. In this case, the diagonal elements of D are the corresponding eigenvalues and $D = P^{-1}AP$, where P is the matrix whose columns are the eigenvectors.

Suppose a matrix A can be diagonalized as above, say $P^{-1}AP = D$ where D is diagonal. Then A has the extremely useful *diagonal factorization* $A = PDP^{-1}$. Using this factorization, the algebra of A reduces to the algebra of the diagonal matrix D, which can be easily calculated. Specifically, suppose $D = \text{diag}(k_1, k_2, \ldots, k_n)$. Then

$$A^m = (PDP^{-1})^m = PD^m P^{-1} = P\text{diag}(k_1^m, k_2^m, \ldots k_n^m)P^{-1}$$

More generally, for any polynomial $f(t)$,

$$f(A) = f(PDP^{-1}) = Pf(D)P^{-1} = P\text{diag}(f(k_1), f(k_2), \ldots f(k_n))P^{-1}$$

Furthermore, if the diagonal entries of D are nonnegative, let

$$B = P\text{diag}(\sqrt{k_1}, \sqrt{k_2}, \ldots, \sqrt{k_n})P^{-1}$$

Then B is a *nonnegative square root* of A; that is, $B^2 = A$ and the eigenvalues of B are nonnegative.

Example 7.5. Let $A = \begin{bmatrix} 3 & 1 \\ 2 & 2 \end{bmatrix}$ and let $v_1 = \begin{bmatrix} 1 \\ -2 \end{bmatrix}$ and $v_2 = \begin{bmatrix} 1 \\ 1 \end{bmatrix}$. Then

$$Av_1 = \begin{bmatrix} 3 & 1 \\ 2 & 2 \end{bmatrix}\begin{bmatrix} 1 \\ -2 \end{bmatrix} = \begin{bmatrix} 1 \\ -2 \end{bmatrix} = v_1 \text{ and } Av_2 = \begin{bmatrix} 3 & 1 \\ 2 & 2 \end{bmatrix}\begin{bmatrix} 1 \\ 1 \end{bmatrix} = \begin{bmatrix} 4 \\ 4 \end{bmatrix} = 4v_2$$

Thus v_1 and v_2 are eigenvectors of A belonging, respectively, to the eigenvalues $\lambda_1 = 1$ and $\lambda_2 = 4$. Observe that v_1 and v_2 are linearly independent and hence form a basis of $\mathbf{R}^2$. Accordingly, A is diagonalizable. Furthermore, let P be the matrix whose columns are the eigenvectors v_1 and v_2. That is, let

$$P = \begin{bmatrix} 1 & 1 \\ -2 & 1 \end{bmatrix}, \text{ and so } P^{-1} = \begin{bmatrix} \frac{1}{3} & -\frac{1}{3} \\ \frac{2}{3} & \frac{1}{3} \end{bmatrix}.$$

Then A is similar to the diagonal matrix

$$D = P^{-1}AP = \begin{bmatrix} \frac{1}{3} & -\frac{1}{3} \\ \frac{2}{3} & \frac{1}{3} \end{bmatrix} \begin{bmatrix} 3 & 1 \\ 2 & 2 \end{bmatrix} \begin{bmatrix} 1 & 1 \\ -2 & 1 \end{bmatrix} = \begin{bmatrix} 1 & 0 \\ 0 & 4 \end{bmatrix}$$

As expected, the diagonal elements 1 and 4 in D are the eigenvalues corresponding, respectively, to the eigenvectors v_1 and v_2, which are the columns of P. In particular, A has the factorization

$$A = PDP^{-1} = \begin{bmatrix} 1 & 1 \\ -2 & 1 \end{bmatrix} \begin{bmatrix} 1 & 0 \\ 0 & 4 \end{bmatrix} \begin{bmatrix} \frac{1}{3} & -\frac{1}{3} \\ \frac{2}{3} & \frac{1}{3} \end{bmatrix} = \begin{bmatrix} 3 & 1 \\ 2 & 2 \end{bmatrix}$$

Accordingly,

$$A^4 = \begin{bmatrix} 1 & 1 \\ -2 & 1 \end{bmatrix} \begin{bmatrix} 1 & 0 \\ 0 & 256 \end{bmatrix} \begin{bmatrix} \frac{1}{3} & -\frac{1}{3} \\ \frac{2}{3} & \frac{1}{3} \end{bmatrix} = \begin{bmatrix} 171 & 85 \\ 170 & 86 \end{bmatrix}$$

Moreover, suppose $f(t) = t^3 - 5t^2 + 3t + 6$; hence $f(1) = 5$ and $f(4) = 2$. Then

$$f(A) = Pf(D)P^{-1} = \begin{bmatrix} 1 & 1 \\ -2 & 1 \end{bmatrix} \begin{bmatrix} 5 & 0 \\ 0 & 2 \end{bmatrix} \begin{bmatrix} \frac{1}{3} & -\frac{1}{3} \\ \frac{2}{3} & \frac{1}{3} \end{bmatrix} = \begin{bmatrix} 3 & -1 \\ -2 & 4 \end{bmatrix}$$

Lastly, we obtain a "positive square root" of A. Specifically, using $\sqrt{1} = 1$ and $\sqrt{4} = 2$, we obtain the matrix

$$B = P\sqrt{D}P^{-1} = \begin{bmatrix} 1 & 1 \\ -2 & 1 \end{bmatrix} \begin{bmatrix} 1 & 0 \\ 0 & 2 \end{bmatrix} \begin{bmatrix} \frac{1}{3} & -\frac{1}{3} \\ \frac{2}{3} & \frac{1}{3} \end{bmatrix} = \begin{bmatrix} \frac{5}{3} & \frac{1}{3} \\ \frac{2}{3} & \frac{4}{3} \end{bmatrix}$$

where $B^2 = A$ and where B has positive eigenvalues 1 and 2.

The above example 7.5 indicates the advantages of a diagonal representation (factorization) of a square matrix. In the following theorem, we list properties that help us to find such a representation.

Theorem 7.6: Let A be a square matrix. Then the following are equivalent.

(i) A scalar λ is an eigenvalue of A.
(ii) The matrix $M = A - \lambda I$ is singular.
(iii) The scalar λ is a root of the characteristic polynomial $\Delta(t)$ of the matrix A.

The eigenspace E_λ of an eigenvalue λ is the solution space of the homogeneous system $MX = 0$, where $M = A - \lambda I$, that is, M is obtained by subtracting λ down the diagonal of A.

Some matrices have no eigenvalues and hence no eigenvectors. However, using Theorem 7.6 and the Fundamental Theorem of Algebra (every polynomial over the complex field **C** has a root), we obtain the following result.

Theorem 7.7: Let A be a square matrix over the complex field **C**. The A has at least one eigenvalue.

Theorem 7.8: Suppose $v_1, v_2, \ldots, v_n$ are nonzero eigenvectors of a matrix A belonging to distinct eigenvalues $\lambda_1, \lambda_2, \ldots, \lambda_n$. Then $v_1, v_2, \ldots, v_n$ are linearly independent.

Theorem 7.9: Suppose the characteristic polynomial $\Delta(t)$ of an n-square matrix A is a product of n distinct factors, say

$$\Delta(t) = (t - a_1)(t - a_2) \ldots (t - a_n).$$

Then A is similar to the diagonal matrix $D = \text{diag}(a_1, a_2, \ldots a_n)$.

If λ is an eigenvalue of a matrix A, then the *algebraic multiplicity* of λ is defined to be the multiplicity of λ as a root of the characteristic polynomial of A, while the *geometric multiplicity* of λ is defined to be the dimension of its eigenspace, $\dim E_\lambda$.

Theorem 7.10: The geometric multiplicity of an eigenvalue λ of a matrix A does not exceed its algebraic multiplicity.

Computing Eigenvalues and Eigenvectors; Diagonalizing Matrices

This section gives an algorithm for computing eigenvalues and eigenvectors for a given square matrix A and for determining whether or not a nonsingular matrix P exists such that $P^{-1}AP$ is diagonal.

Algorithm 7.1: (**Diagonalization Algorithm**) The input is an n-square matrix A.

Step 1. Find the characteristic polynomial $\Delta(t)$ of A.
Step 2. Find the roots of $\Delta(t)$ to obtain the eigenvalues of A.
Step 3. Repeat (a) and (b) for each eigenvalue λ of A.

(a) Form the matrix $M = A - \lambda I$ by subtracting λ down the diagonal of A.

(b) Find a basis for the solution space of the homogeneous system $MX = 0$. (These basis vectors are linearly independent eigenvectors of A belonging to λ.)

Step 4. Consider the collection $S = \{v_1, v_2, \dots, v_m\}$ of all eigenvectors obtained in step 3.

(a) If $m \neq n$, then A is not diagonalizable.

(b) If $m = n$, then A is diagonalizable. Specifically, let P be the matrix whose columns are the eigenvectors $v_1, v_2, \dots, v_n$. Then $D = P^{-1}AP = \text{diag}(\lambda_1, \lambda_2, \dots, \lambda_n)$ where λ_i is the eigenvalue corresponding to the eigenvector v_i.

Example 7.6. The diagonalizable algorithm is applied to $A = \begin{bmatrix} 4 & 2 \\ 3 & -1 \end{bmatrix}$.

(1) The characteristic polynomial $\Delta(t)$ of A is computed. We have

$$\text{tr}(A) = 4 - 1 = 3, \quad |A| = -4 - 6 = -10;$$

hence $\Delta(t) = t^2 - 3t - 10 = (t - 5)(t + 2)$

(2) Set $\Delta(t) = (t - 5)(t + 2) = 0$. The roots $\lambda_1 = 5$ and $\lambda_2 = -2$ are the eigenvalues of A.

(3) (a) We find an eigenvector v_1 of A belonging to the eigenvalue $\lambda_1 = 5$. Subtracting $\lambda_1 = 5$ down the diagonal of A to obtain the matrix $M = \begin{bmatrix} -1 & 2 \\ 3 & -6 \end{bmatrix}$. The eigenvectors belonging to $\lambda_1 = 5$ form the solution of the homogeneous system $MX = 0$, that is,

$$\begin{bmatrix} -1 & 2 \\ 3 & -6 \end{bmatrix}\begin{bmatrix} x \\ y \end{bmatrix} = \begin{bmatrix} 0 \\ 0 \end{bmatrix} \text{ or } \begin{matrix} -x + 2y = 0 \\ 3x - 6y = 0 \end{matrix} \text{ or } -x + 2y = 0$$

The system has only one free variable. Thus a nonzero solution, for example, $v_1 = (2, 1)$, is an eigenvector that spans the eigenspace of $\lambda_1 = 5$.

(*b*) We find an eigenvector v_2 of A belonging to the eigenvalue $\lambda_2 = -2$. Subtracting -2 (or add 2) down the diagonal of A to obtain the matrix $M = \begin{bmatrix} 6 & 2 \\ 3 & 1 \end{bmatrix}$ and the homogeneous system

$$\begin{aligned} 6x + 2y &= 0 \\ 3x + y &= 0 \end{aligned} \quad \text{or } 3x + y = 0$$

The system has only one independent solution. Thus a nonzero solution, say, $v_2 = (-1, 3)$, is an eigenvector that spans the eigenspace of $\lambda_2 = -2$.

(4) Let P be the matrix whose columns are the eigenvectors v_1 and v_2. Then

$$P = \begin{bmatrix} 2 & -1 \\ 1 & 3 \end{bmatrix}, \text{ and so } P^{-1} = \begin{bmatrix} \frac{3}{7} & \frac{1}{7} \\ -\frac{1}{7} & \frac{2}{7} \end{bmatrix}$$

Accordingly, $D = P^{-1}AP$ is the diagonal matrix whose diagonal entries are the corresponding eigenvalues; that is,

$$D = P^{-1}AP = \begin{bmatrix} \frac{3}{7} & \frac{1}{7} \\ -\frac{1}{7} & \frac{2}{7} \end{bmatrix} \begin{bmatrix} 4 & 2 \\ 3 & -1 \end{bmatrix} \begin{bmatrix} 2 & -1 \\ 1 & 3 \end{bmatrix} = \begin{bmatrix} 5 & 0 \\ 0 & -2 \end{bmatrix}$$

Example 7.7. Consider the matrix $B = \begin{bmatrix} 5 & -1 \\ 1 & 3 \end{bmatrix}$. We have

$$\operatorname{tr}(B) = 5 + 3 = 8, \quad |B| = 15 + 1 = 16; \quad \text{so} \quad \Delta(t) = t^2 - 8t + 16 = (t-4)^2$$

Accordingly, $\lambda = 4$ is the only eigenvalue of B. Subtract $\lambda = 4$ down the diagonal of B to obtain the matrix $M = \begin{bmatrix} 1 & -1 \\ 1 & -1 \end{bmatrix}$ and the homogeneous system $\begin{aligned} x - y &= 0 \\ x - y &= 0 \end{aligned}$ or $x - y = 0$.

The system has only one independent solution; say, $x = 1$, $y = 1$. Thus $v = (1, 1)$ and its multiples are the only eigenvectors of B. Accordingly, B is not diagonalizable, since there does not exist a basis consisting of eigenvectors of B.

Example 7.8. Consider the matrix $A = \begin{bmatrix} 3 & -5 \\ 2 & -3 \end{bmatrix}$. Here tr($A$) = 3 – 3 = 0 and $|A| = -9 + 10 = 1$. Thus $\Delta(t) = t^2 + 1$ is the characteristic polynomial of A. We consider two cases:

(*a*) A is a matrix over the real field **R**. Then $\Delta(t)$ has no (real) roots. Thus A has no eigenvalues and no eigenvectors, and so A is not diagonalizable.

(*b*) A is a matrix over the complex field **C**. Then $\Delta(t) = (t - i)(t + i)$ has two roots, i and $-i$. Thus A has two distinct eigenvalues i and $-i$, and hence A has two independent eigenvectors. Accordingly there exists a nonsingular matrix P over the complex field **C** for which $P^{-1}AP = \begin{bmatrix} i & 0 \\ 0 & -i \end{bmatrix}$. Therefore A is diagonalizable (over **C**).

There are many real matrices that are not diagonalizable. In fact, some real matrices may not have any (real) eigenvalues. However, if A is a real symmetric matrix, then these problems do not exist. Namely, we have the following theorems.

Theorem 7.11: Let A be a real symmetric matrix. Then each root λ of its characteristic polynomial is real.

Theorem 7.12: Let A be a real symmetric matrix. Suppose u and v are eigenvectors of A belonging to distinct eigenvalues λ_1 and λ_2. Then u and v are orthogonal, that is, $\langle u, v \rangle = 0$.

The above two theorems give us the following fundamental result.

Theorem 7.13: Let A be a real symmetric matrix. Then there exists an orthogonal matrix P such that $D = P^{-1}AP$ is diagonal.

The orthogonal matrix P is obtained by normalizing a basis of orthogonal eigenvectors of A as illustrated below. In such a case, we say that A is "orthogonally diagonalizable."

Example 7.9. Let $A = \begin{bmatrix} 2 & -2 \\ -2 & 5 \end{bmatrix}$, a real symmetric matrix. Find an orthogonal matrix P such that $P^{-1}AP$ is diagonal.

(*a*) First we find the characteristic polynomial $\Delta(t)$ of A. We have $\operatorname{tr}(A) = 2 + 5 = 7$, $|A| = 10 - 4 = 6$; so

$$\Delta(t) = t^2 - 7t + 6 = (t - 6)(t - 1).$$

Accordingly, $\lambda_1 = 6$ and $\lambda_2 = 1$ are the eigenvalues of A. Subtracting $\lambda_1 = 6$ down the diagonal of A yields the matrix $M = \begin{bmatrix} -4 & -2 \\ -2 & -1 \end{bmatrix}$ and the homogeneous system $\begin{matrix} -4x - 2y = 0 \\ -2x - y = 0 \end{matrix}$ or $2x + y = 0$. A nonzero solution is $u_1 = (1, -2)$.

(*b*) Subtracting $\lambda_2 = 1$ down the diagonal of A yields the matrix $M = \begin{bmatrix} 1 & -2 \\ -2 & 4 \end{bmatrix}$ and the homogeneous system $x - 2y = 0$. A nonzero solution is $u_2 = (2, 1)$.

As expected from Theorem 7.12, u_1 and u_2 are orthogonal. Normalizing u_1 and u_2 yields the orthonormal vectors

$$\hat{u}_1 = \left(1/\sqrt{5}, -2/\sqrt{5}\right) \quad \text{and} \quad \hat{u}_2 = \left(2/\sqrt{5}, 1/\sqrt{5}\right)$$

Finally, let P be the matrix whose columns are u_1 and u_2, respectively. Then $P = \begin{bmatrix} 1/\sqrt{5} & 2/\sqrt{5} \\ -2/\sqrt{5} & 1/\sqrt{5} \end{bmatrix}$ and $P^{-1}AP = \begin{bmatrix} 6 & 0 \\ 0 & 1 \end{bmatrix}$. As expected, the diagonal entries of $P^{-1}AP$ are the eigenvalues corresponding to the columns of P.

The procedure in the above Example 7.9 is formalized in the following algorithm, which finds an orthogonal matrix P such that $P^{-1}AP$ is diagonal.

Algorithm 7.2: (**Orthogonal Diagonalization Algorithm**) The input is a real symmetric matrix A.

Step 1. Find the characteristic polynomial $\Delta(t)$ of A.
Step 2. Find the eigenvalues of A, which are the roots of $\Delta(t)$.
Step 3. For each eigenvalue λ of A in Step 2, find an orthogonal basis of its eigenspace.
Step 4. Normalize all eigenvectors in Step 3, which then forms an orthonormal basis of $\mathbf{R}^n$.

Step 5. Let P be the matrix whose columns are the normalized eigenvectors in Step 4.

Minimal Polynomials

Let A be any square matrix. Let $J(A)$ denote the collection of all polynomials $f(t)$ for which A is a root, i.e., for which $f(A) = 0$. The set $J(A)$ is not empty, since the Cayley-Hamilton Theorem 7.2 tells us that the characteristic polynomial $\Delta_A(t)$ of A belongs to $J(A)$. Let $m(t)$ denote the monic polynomial of lowest degree in $J(A)$. (Such a polynomial $m(t)$ exists and is unique.) We call $m(t)$ the *minimal polynomial* of the matrix A.

A polynomial $f(t) \neq 0$ is *monic* if its leading coefficient equals one.

Theorem 7.14: The minimal polynomial $m(t)$ of a matrix A divides every polynomial that has A as a zero. In particular, $m(t)$ divides the characteristic polynomial $\Delta(t)$ of A.

There is an even stronger relationship between $m(t)$ and $\Delta(t)$.

Theorem 7.15: The characteristic polynomial $\Delta(t)$ and the minimal polynomial $m(t)$ of a matrix A have the same irreducible factors.

This theorem does not say that $m(t) = \Delta(t)$, only that any irreducible factor of one must divide the other. In particular, since a linear factor is irreducible, $m(t)$ and $\Delta(t)$ have the same linear factors. Hence they have the same roots. Thus we have the following theorem.

Theorem 7.16: A scalar λ is an eigenvalue of the matrix A if and only if λ is a root of the minimal polynomial of A.

Example 7.10. Find the minimal polynomial $m(t)$ of $A = \begin{bmatrix} 2 & 2 & -5 \\ 3 & 7 & -15 \\ 1 & 2 & -4 \end{bmatrix}$.

First find the characteristic polynomial $\Delta(t)$ of A. We have $\mathrm{tr}(A) = 5$, $A_{11} + A_{22} + A_{33} = 2 - 3 + 8 = 7$ and $|A| = 3$. Hence

$$\Delta(t) = t^3 - 5t^2 + 7t - 3 = (t-1)^2(t-3).$$

The minimal polynomial $m(t)$ must divide $\Delta(t)$. Also, each irreducible factor of $\Delta(t)$, that is, $t - 1$ and $t - 3$, must also be a factor of $m(t)$. Thus $m(t)$ is exactly one of the following:

$$f(t) = (t-3)(t-1) \text{ or } g(t) = (t-3)(t-1)^2.$$

We know, by the Cayley-Hamilton Theorem, that $g(A) = \Delta(A) = 0$. Hence we need only test $f(t)$. We have

$$f(A) = (A - I)(A - 3I) = \begin{bmatrix} 1 & 2 & -5 \\ 3 & 6 & -15 \\ 1 & 2 & -5 \end{bmatrix} \begin{bmatrix} -1 & 2 & -5 \\ 3 & 4 & -15 \\ 1 & 2 & -7 \end{bmatrix} = \begin{bmatrix} 0 & 0 & 0 \\ 0 & 0 & 0 \\ 0 & 0 & 0 \end{bmatrix}$$

Thus $f(t) = m(t) = (t-1)(t-3) = t^2 - 4t + 3$ is the minimal polynomial of A.

Example 7.11.

(*a*) Consider the following two r-square matrices, where $a \neq 0$:

$$J(\lambda, r) = \begin{bmatrix} \lambda & 1 & 0 & \dots & 0 & 0 \\ 0 & \lambda & 1 & \dots & 0 & 0 \\ \dots & \dots & \dots & \dots & \dots & \dots \\ 0 & 0 & 0 & \dots & \lambda & 1 \\ 0 & 0 & 0 & \dots & 0 & \lambda \end{bmatrix} \text{ and } A = \begin{bmatrix} \lambda & a & 0 & \dots & 0 & 0 \\ 0 & \lambda & a & \dots & 0 & 0 \\ \dots & \dots & \dots & \dots & \dots & \dots \\ 0 & 0 & 0 & \dots & \lambda & a \\ 0 & 0 & 0 & \dots & 0 & \lambda \end{bmatrix}$$

The first matrix, called a Jordan Block, has λ's on the diagonal, 1's on the *superdiagonal* (consisting of the entries above the diagonal entries), and 0's elsewhere. The second matrix A has λ's on the diagonal, *a*s on the *superdiagonal*, and 0's elsewhere. [Thus A is a generalization of $J(\lambda, r)$.] One can show that $f(t) = (t - \lambda)^r$ is both the characteristic and minimal polynomial of both $J(\lambda, r)$ and A.

(*b*) Consider an arbitrary monic polynomial

$$f(t) = t^n + a_{n-1}t^{n-1} + \ldots + a_1 t + a_0$$

Let $C(f)$ be the n-square matrix with 1's on the *subdiagonal* (consisting of the entries below the diagonal entries), the negatives of the coefficients in the last column, and 0's elsewhere as follows:

$$C(f) = \begin{bmatrix} 0 & 0 & \ldots & 0 & -a_0 \\ 1 & 0 & \ldots & 0 & -a_1 \\ 0 & 1 & \ldots & 0 & -a_2 \\ \ldots & \ldots & \ldots & \ldots & \ldots \\ 0 & 0 & \ldots & 1 & -a_{n-1} \end{bmatrix}$$

Then $C(f)$ is called the *companion matrix* of the polynomial $f(t)$. Moreover, the minimal polynomial $m(t)$ and the characteristic polynomial $\Delta(t)$ of the companion matrix $C(f)$ are both equal to the original polynomial $f(t)$.

Suppose M is a block triangular matrix, say $M = \begin{bmatrix} A_1 & B \\ 0 & A_2 \end{bmatrix}$ where A_1 and A_2 are square matrices. Then $tI - M$ is also a block triangular matrix, with diagonal blocks $tI - A_1$ and $tI - A_2$. Thus

$$|tI - M| = \begin{vmatrix} tI - A_1 & -B \\ 0 & tI - A_2 \end{vmatrix} = |tI - A_1||tI - A_2|$$

That is, the characteristic polynomial of M is the product of the characteristic polynomials of the diagonal blocks A_1 and A_2.

By induction, we obtain the following useful result.

Theorem 7.17: Suppose M is a block triangular matrix with diagonal blocks $A_1, A_2, \ldots, A_r$. Then the characteristic polynomial of M is the product of the characteristic polynomials of the diagonal blocks A_i; that is,

$$\Delta_M(t) = \Delta_{A_1}(t)\Delta_{A_2}(t)\ldots\Delta_{A_r}(t)$$

Example 7.12. Consider the matrix $M = \left[\begin{array}{cc|cc} 9 & -1 & 5 & 7 \\ 8 & 3 & 2 & -4 \\ \hline 0 & 0 & 3 & 6 \\ 0 & 0 & -1 & 8 \end{array}\right]$. Then M is a block triangular matrix with diagonal blocks

$$A = \begin{bmatrix} 9 & -1 \\ 8 & 3 \end{bmatrix} \text{ and } B = \begin{bmatrix} 3 & 6 \\ -1 & 8 \end{bmatrix}$$

$$\begin{aligned} \operatorname{tr}(A) = 9 + 3 = 12, \quad \det(A) = 27 + 8 = 35, \quad \text{so} \quad \Delta_A(t) &= t^2 - 12t + 35 \\ &= (t-5)(t-7) \\ \operatorname{tr}(B) = 3 + 8 = 11, \quad \det(B) = 24 + 6 = 30, \quad \text{so} \quad \Delta_B(t) &= t^2 - 11t + 30 \\ &= (t-5)(t-6) \end{aligned}$$

Accordingly, the characteristic polynomial of M is the product

$$\Delta_M(t) = \Delta_A(t)\,\Delta_B(t) = (t-5)^2(t-6)(t-7)$$

Theorem 7.18: Suppose M is a block diagonal matrix with diagonal blocks $A_1, A_2, \ldots, A_r$. Then the minimal polynomial of M is equal to the least common multiple (LCM) of the minimal polynomials of the diagonal blocks A_i.

We emphasize that this theorem applies to block diagonal matrices, whereas the analogous Theorem 7.17 on characteristic polynomials applies to block triangular matrices.

Example 7.13. Find the characteristic polynomial $\Delta(t)$ and the minimal polynomial $m(t)$ of the block diagonal matrix

$$M = \left[\begin{array}{cc|cc|c} 2 & 5 & 0 & 0 & 0 \\ 0 & 2 & 0 & 0 & 0 \\ \hline 0 & 0 & 4 & 2 & 0 \\ 0 & 0 & 3 & 5 & 0 \\ \hline 0 & 0 & 0 & 0 & 7 \end{array}\right] = \operatorname{diag}(A_1, A_2, A_3),$$

where

$$A_1 = \begin{bmatrix} 2 & 5 \\ 0 & 2 \end{bmatrix}, A_2 = \begin{bmatrix} 4 & 2 \\ 3 & 5 \end{bmatrix}, A_3 = [7]$$

Then $\Delta(t)$ is the product of the characteristic polynomials $\Delta_1(t)$, $\Delta_2(t)$, $\Delta_3(t)$ of A_1, A_2, A_3, respectively. One can show that

$$\Delta_1(t) = (t-2)^2, \quad \Delta_2(t) = (t-2)(t-7), \quad \Delta_3(t) = t-7.$$

Thus $\Delta(t) = (t-2)^3(t-7)^2$.

The minimal polynomials $m_1(t)$, $m_2(t)$, $m_3(t)$ of the diagonal blocks A_1, A_2, A_3, respectively, are equal to the characteristic polynomials, that is,

$$m_1(t) = (t-2)^2, m_2(t) = (t-2)(t-7), m_3(t) = t-7.$$

But $m(t)$ is equal to the least common multiple of $m_1(t)$, $m_2(t)$, $m_3(t)$.

Thus $m(t) = (t-2)^2(t-7)$.

Chapter 8
LINEAR MAPPINGS

IN THIS CHAPTER:

- ✔ *Mappings; Functions*
- ✔ *Linear Mappings*
- ✔ *Kernel and Image of a Linear Mapping*
- ✔ *Operations with Linear Mappings*
- ✔ *Linear Mappings and Matrices*
- ✔ *Change of Basis*
- ✔ *Matrices and General Linear Mappings*

Mappings; Functions

The main subject matter of linear algebra is the study of linear mappings and their representation by means of matrices. This chapter introduces us to these linear maps and shows how they can be represented by matrices. First, however, we begin with a study of mappings in general.

Let A and B be arbitrary nonempty sets. Suppose to each element in A there is assigned a unique element of B; the collection f of such assignments is called a *mapping* (or *map*) from A into B, and is denoted by

$f: A \to B$. The set A is called the *domain* of the mapping, and B is called the *target set*. We write $f(a)$, read "f of a," for the element of B that f assigns to $a \in A$.

One may also view a mapping $f: A \to B$ as a computer that, for each input value $a \in A$, produces a unique output $f(a) \in B$. The term *function* is used synonymously with the word mapping, although some texts reserve the word "function" for a real-valued mapping.

Consider a mapping $f: A \to B$. If A' is any subset of A, then $f(A')$ denotes the set of images of elements of A'; and if B' is any subset of B, then $f^{-1}(B')$ denotes the set of elements of A, each of whose image lies in B. That is, $f(A') = \{f(a): a \in A'\}$ and $f^{-1}(B') = \{a \in A: f(a) \in B'\}$. We call $f(A')$ the *image* of A' and $f^{-1}(B')$ the *inverse image* or *preimage* of B'. In particular, the set of all images, i.e., $f(A)$, is called the image or *range* of f.

To each mapping $f: A \to B$ there corresponds the subset of $A \times B$ given by $\{(a, f(a)) : a \in A\}$. We call this set the *graph* of f. Two mappings $f: A \to B$ and $g: A \to B$ are defined to be *equal*, written $f = g$, if $f(a) = g(a)$ for every $a \in A$, that is, if they have the same graph. Thus we do not distinguish between a function and its graph. The negation of $f = g$ is written $f \neq g$ and is the statement:

There exists an $a \in A$ for which $f(a) \neq g(a)$.

Sometimes the "barred" arrow $\mapsto$ is used to denote the image of an arbitrary element $a \in A$ under a mapping $f: A \to B$ by writing $x \mapsto f(x)$. This is illustrated in the following example.

Example 8.1.

(*a*) Let $f: \mathbf{R} \to \mathbf{R}$ be the function that assigns to each real number x its square x^2. We can denote this function by writing $f(x) = x^2$ or $x \mapsto x^2$. Here the image of -3 is 9, so we may write $f(-3) = 9$. However, $f^{-1}(9) = \{-3, 3\}$. Also, $f(\mathbf{R}) = [0,\infty) = \{x : x \geq 0\}$ is the image of f.

(*b*) Let $A = \{a, b, c, d\}$ and $B = \{x, y, z, t\}$. Then the following defines a mapping $f: A \to B$:

$$f(a) = y, f(b) = x, f(c) = z, f(d) = y \text{ or } f = \{(a, y), (b, x), (c, z), (d, y)\}$$

The first defines the mapping explicitly, and the second defines the mapping by its graph. Here,

$$f(\{a, b, d\}) = \{f(a), f(b), f(d)\} = \{y, x, y\} = \{x, y\}$$

Furthermore, $f(A) = \{x, y, z\}$ is the image of f.

Example 8.2. Let V be the vector space of polynomials over **R**, and let $p(t) = 3t^2 - 5t + 2$.

(*a*) The derivative defines a mapping $\mathbf{D}: V \to V$ where, for any polynomials $f(t)$, we have $\mathbf{D}(f) = df/dt$. Thus

$$\mathbf{D}(p) = \mathbf{D}(3t^2 - 5t + 2) = 6t - 5$$

(*b*) The integral, say from 0 to 1, defines a mapping $\boldsymbol{J}: V \to \mathrm{R}$. That is, for any polynomial $f(t)$,

$$J(f) = \int_0^1 f(t)dt, \text{ and so } J(p) = \int_0^1 (3t^2 - 5t + 2)dt = \tfrac{1}{2}$$

Observe that the mapping in (*b*) is from the vector space V into the scalar field **R**, whereas the mapping in (*a*) is from the vector space V into itself.

Let A be any $m \times n$ matrix over K. Then A determines a mapping F_A: $K^n \to K^m$ by $F_A(u) = Au$ where the vectors in K^n and K^m are written as columns. For example, suppose

$$A = \begin{bmatrix} 1 & -4 & 5 \\ 2 & 3 & -6 \end{bmatrix} \text{ and } u = \begin{bmatrix} 1 \\ 3 \\ -5 \end{bmatrix}$$

then

$$F_A(u) = Au = \begin{bmatrix} 1 & -4 & 5 \\ 2 & 3 & -6 \end{bmatrix} \begin{bmatrix} 1 \\ 3 \\ -5 \end{bmatrix} = \begin{bmatrix} -36 \\ 41 \end{bmatrix}$$

For notational convenience, we shall frequently denote the mapping F_A by the letter A, the same symbol as used for the matrix.

Consider two mappings $f: A \rightarrow B$ and $g: B \rightarrow C$, illustrated below

$$A \xrightarrow{f} B \xrightarrow{g} C$$

The composition of f and g, denoted $g \circ f$, is the mapping $g \circ f: A \rightarrow C$ defined by $(g \circ f)(a) = g(f(a))$. That is, first we apply f to $a \in A$ to get the output $f(a) \in B$ using f, and then we input $f(a)$ to get the output $g(f(a)) \in C$ using g.

Our first theorem tells us that the composition of mappings satisfies the associative law.

Theorem 8.1: Let $f : A \rightarrow B$, $g : B \rightarrow C$, $h : C \rightarrow D$. Then

$$h \circ (g \circ f) = (h \circ g) \circ f$$

We prove this theorem here. Let $a \in A$. Then

$$(h \circ (g \circ f))(a) = h((g \circ f)(a)) = h(g(f(a)))$$
$$((h \circ g) \circ f)(a) = (h \circ g) f(a)) = h(g(f(a)))$$

Thus($h \circ (g \circ f))(a) = ((h \circ g) \circ f)(a)$ for every $a \in A$, and so

$$h \circ (g \circ f) = (h \circ g) \circ f.$$

A mapping $f : A \rightarrow B$ is said to be *one-to-one* (or 1-1 or *injective*) if different elements of A have distinct images; that is:

(1) If $a \neq a'$, then $f(a)$ $(f(a')$

or equivalently

(2) if $f(a) = f(a')$, then $a = a'$.

A mapping $f : A \rightarrow B$ is said to be *onto* (or f maps A onto B or *surjective*) if every $b \in B$ is the image of at least one $a \in A$.

A mapping $f : A \rightarrow B$ is said to be a *one-to-one correspondance* between A and B (or *bijective*) if f is both one-to-one and onto.

Example 8.3. Let $f : \mathbf{R} \to \mathbf{R}$, $g : \mathbf{R} \to \mathbf{R}$, be defined by

$$f(x) = 2^x,\ g(x) = x^3 - x$$

The function f is one-to-one since $2^a = 2^b$ implies $a = b$ but f is not onto since, for example, -2 has no preimage. The function g is onto however it is not one-to-one since $g(0) = g(1) = 0$.

Let A be any nonempty set. The mapping $f : A \to A$ defined by $f(a) = a$, that is, the function that assigns to each element in A itself, is called the *identity mapping*. It is usually denoted by $\mathbf{1}_A$ or $\mathbf{1}$ or I. We emphasize that f has an inverse if and only if f is a one-to-one correspondence between A and B, that is, f is one-to-one and onto. Also, if $b \in B$, then $f^{-1}(b) = a$, where a is the unique element of A for which $f(a) = b$.

Linear Mappings

Let V and U be vector spaces over the same field K. A mapping $F : V \to U$ is called a *linear mapping* or *linear transformation* if it satisfies the following two conditions:

(1) For any vectors $v, w \in V$, $F(v + w) = F(v) + F(w)$.

(2) For any scalar k and vector $v \in V$, $F(kv) = kF(v)$.

Namely, $F : V \to U$ is linear if it "preserves" the two basic operations of a vector space, that of vector addition and that of scalar multiplication.

Substituting $k = 0$ into condition (2), we obtain $F(0) = 0$. Thus, every linear mapping takes the zero vector into the zero vector.

Now for any scalars $a, b \in K$ and any vectors $v, w \in V$, we obtain

$$F(av + bw) = F(av) + F(bw) = aF(v) + bF(w)$$

More generally, for any scalars $a_i \in K$ and any vector $v_i \in V$, we obtain the following basic property of linear mappings:

$$F(a_1v_1 + a_2v_2 + \dots + a_mv_m) = a_1F(v_1) + a_2F(v_2) + \dots + a_mF(v_m)$$

A linear mapping $F(av + bw) = aF(v) + bF(w)$ and so this condition is sometimes used as its definition.

The term *linear transformation* rather than linear mapping is frequently used for linear mappings of the form $F : \mathbf{R}^n \to \mathbf{R}^m$.

Example 8.4.

(*a*) Let $F : \mathbf{R}^3 \to \mathbf{R}^3$ be the "projection" mapping into the xy-plane, that is, F is the mapping defined by $F(x,y,z) = (x, y, 0)$. We show that F is linear. Let $v = (a, b, c)$ and $w = (a', b', c')$. Then

$$F(v + w) = F(a + a', b + b', c + c') = (a + a', b + b', 0)$$
$$= (a, b, 0) + (a', b', 0) = F(v) + F(w)$$

and, for any scalar k,

$$F(kv) = F((ka, kb, kc) = (ka, kb, 0) = k(a, b, 0) = kF(v).$$

Thus F is linear.

(*b*) Let $G : \mathbf{R}^2 \to \mathbf{R}^2$ be the "translation" mapping defined by $G(x, y) = (x + 1, y + 2)$. [That is, G adds the vector $(1, 2)$ to any vector $v = (x, y)$ in $\mathbf{R}^2$.] Note that $G(0) = G(0, 0) = (1, 2) \neq 0$. Thus the zero vector is not mapped into the zero vector. Hence G is not linear.

Example 8.5. Consider the vector space $V = \mathbf{P}(t)$ of polynomials over the real field $\mathbf{R}$. Let $u(t)$ and $v(t)$ be any polynomials in V and let k be any scalar.

(*a*) Let $\mathbf{D}: V \to V$ be the derivative mapping. One proves in calculus that

$$\frac{d(u+v)}{dt} = \frac{du}{dt} + \frac{dv}{dt} \text{ and } \frac{d(ku)}{dt} = k\frac{du}{dt}$$

That is, $\mathbf{D}(u + v) = \mathbf{D}(u) + \mathbf{D}(v)$ and $\mathbf{D}(ku) = k\mathbf{D}(u)$. Thus the derivative mapping is linear.

(*b*) Let $\mathbf{J} : V \to \mathbf{R}$ be an integral mapping, say $J(f(t)) = \int_0^1 f(t)dt$. One also proves in calculus that,

$$\int_0^1 [u(t) + v(t)]dt = \int_0^1 u(t)dt + \int_0^1 v(t)dt$$

and

$$\int_0^1 ku(t)dt = k\int_0^1 u(t)dt$$

That is, $\mathbf{J}(u + v) = \mathbf{J}(u) + \mathbf{J}(v)$ and $\mathbf{J}(ku) = k\mathbf{J}(u)$. Thus the integral mapping is linear.

Example 8.6.

(*a*) Let $F : V \to U$ be the mapping that assigns the zero vector $0 \in U$ to every vector $v \in V$. Then, for any vectors $v, w \in V$ and any scalar $k \in K$, we have

$$F(v + w) = 0 = 0 + 0 = F(v) + F(w) \text{ and } F(kv) = 0 = k0 = kF(v)$$

Thus F is linear. We call F the *zero mapping*, and we shall usually denote it by 0.

(*b*) Consider the identity mapping $I : V \to V$, which maps each $v \in V$ into itself. Then, for any vectors $v, w \in V$ and any scalars $a, b \in K$, we have $I(av + bw) = av + bw = aI(v) + bI(w)$. Thus I is linear.

Our next theorem gives us an abundance of examples of linear mappings. In particular, it tells us that a linear mapping is completely determined by its values on the elements of a basis.

Theorem 8.2: Let V and U be vector spaces over a field K. Let $\{v_1, v_2, \ldots, v_n\}$ be a basis of V and let $u_1, u_2, \ldots, u_n$ be any vectors in U. Then there exists a unique linear mapping $F : V \to U$ such that

$$F(v_1) = u_1, F(v_2) = u_2, \ldots, F(v_n) = u_n.$$

We emphasize that the vectors $u_1, u_2, \ldots, u_n$ in Theorem 8.2 are completely arbitrary; they may be linearly dependent or they may even be equal to each other.

Let A be any real $m \times n$ matrix. Recall that A determines a mapping $F_A: K^n \to K^m$ by $F_A(u) = Au$ (where the vectors in K^n and K^m are written as columns). We show F_A is linear. By matrix multiplication,

$$F_A(v + w) = A(v + w) = Av + Aw = F_A(v) + F_A(w)$$
$$F_A(kv) = A(kv) = k(Av) = kF_A(v)$$

In other words, using A to represent the mapping, we have

$$A(v + w) = Av + Aw$$
$$A(kv) = k(Av)$$

Thus the matrix mapping A is linear.

Two vector spaces V and U over K are *isomorphic*, written $V \cong U$, if there exists a bijective (one-to-one and onto) linear mapping $F : V \to U$. The mapping F is then called an *isomorphism* between V and U.

Consider any vector space V of dimension n and let S be any basis of V. Then the mapping $v \mapsto [v]_s$ which maps each vector $v \in V$ into its coordinate vector $[v]_s$, is an isomorphism between V and K^n.

Kernel and Image of a Linear Mapping

Let $F : V \to U$ be a linear mapping. The *kernel* of F, written Ker F, is the set of elements in V that map into the zero vector 0 in U; that is,

$$\text{Ker } F = \{v \in V : F(v) = 0\}$$

The *image* (or *range*) of F, written Im F, is the set of image points in U; that is,

$$\text{Im } F = \{u \in U: \text{there exists } v \in V \text{ for which } F(v) = u\}$$

Theorem 8.3: Let $F : V \to U$ be a linear mapping. Then the kernel of F is a subspace of V and the image of F is a subspace of U.

Now suppose that $v_1, v_2, \ldots, v_m$ span a vector space V and that $F : V \to U$ is linear. We show that $F(v_1), F(v_2), \ldots, F(v_m)$ span Im F. Let $u \in$ Im F. Then there exists $v \in V$ such that $F(v) = u$. Since the v's span V and since $v \in V$, there exist scalars $a_1, a_2, \ldots, a_m$ for which $v = a_1v_1 + a_2v_2 + \ldots + a_mv_m$. Therefore,

$$u = F(v) = F(a_1v_1 + a_2v_2 + \dots + a_mv_m)$$
$$= a_1F(v_1) + a_2F(v_2) + \dots + a_mF(v_m).$$

Thus the vectors $F(v_1), F(v_2), \dots, F(v_m)$ span Im F.

We formally state the above result.

Proposition 8.4: Suppose $v_1, v_2, \dots, v_m$ span a vector space V, and suppose $F : V \to U$ is linear. Then $F(v_1), F(v_2), \dots, F(v_m)$ span Im F.

Example 8.7. Let F: $\mathbf{R}^3 \to \mathbf{R}^3$ be the projection of a vector v into the xy-plane; that is $F(x, y, z) = (x, y, 0)$. Clearly the image of F is the entire xy-plane, i.e., points of the form $(x, y, 0)$. Moreover, the kernel of F is the z-axis, i.e., points of the form $(0, 0, c)$. That is,

$$\text{Im } F = \{(a, b, c): c = 0\} = xy\text{-plane}$$

and

$$\text{Ker } F = \{(a, b, c): a = 0, b = 0\} = z\text{-axis}.$$

Example 8.8. Consider the vector space $V = \mathbf{P}(t)$ of polynomials over the real field $\mathbf{R}$, and let H: $V \to V$ be the third-derivative operator, that is, $H[f(t)] = d^3f / dt^3$. [Sometimes the notation $\mathbf{D}^3$ is used for H, where $\mathbf{D}$ is the derivative operator.] We claim that

$$\text{Ker } H = \{\text{polynomials of degree} \le 2\} = \mathbf{P}_2(t) \text{ and Im } H = V.$$

The first comes from the fact that $H(at^2 + bt + c) = 0$ but $H(t^n) \neq 0$ for $n \ge 3$. The second comes from that fact that every polynomial $g(t)$ in V is the third derivative of some polynomial $f(t)$ (which can be obtained by taking the antiderivative of $g(t)$ three times).

Consider, say, a 3×4 matrix A and the usual basis $\{e_1, e_2, e_3, e_4\}$ of K^4 written as columns):

$$A = \begin{bmatrix} a_1 & a_2 & a_3 & a_4 \\ b_1 & b_2 & b_3 & b_4 \\ c_1 & c_2 & c_3 & c_4 \end{bmatrix},$$

$$e_1 = \begin{bmatrix} 1 \\ 0 \\ 0 \\ 0 \end{bmatrix}, \quad e_2 = \begin{bmatrix} 0 \\ 1 \\ 0 \\ 0 \end{bmatrix}, \quad e_3 = \begin{bmatrix} 0 \\ 0 \\ 1 \\ 0 \end{bmatrix}, \quad e_4 = \begin{bmatrix} 0 \\ 0 \\ 0 \\ 1 \end{bmatrix}$$

Recall that A may be viewed as a linear mapping $A: K^4 \to K^3$, where the vectors in K^4 and K^3 are viewed as column vectors. Now the usual basis vectors span K^4, so their images Ae_1, Ae_2, Ae_3, Ae_4 span the image of A. But the vectors Ae_1, Ae_2, Ae_3, Ae_4 are precisely the columns of A:

$$Ae_1 = [a_1, b_1, c_1]^T, Ae_2 = [a_2, b_2, c_2]^T,$$
$$Ae_3 = [a_3, b_3, c_3]^T, Ae_4 = [a_4, b_4, c_4]^T,$$

Thus the image of A is precisely the column space of A.

On the other hand, the kernel of A consists of all vectors v for which $Av = 0$. This means that the kernel of A is the solution space of the homogeneous system $AX = 0$, called the *null space* of A.

We state the above results formally.

Proposition 8.5: Let A be any $m \times n$ matrix over a field K viewed as a linear map $A: K^n \to K^m$. Then $\text{Ker } A = \text{nullsp}(A)$ and $\text{Im } A = \text{colsp}(A)$. Here colsp($A$) denotes the column space of A, and nullsp(A) denotes the null space of A.

Let $F: V \to U$ be a linear mapping. The *rank* of F is defined to be the dimension of its image, and the *nullity* of F is defined to be the dimension of its kernel; namely,

$$\text{rank}(F) = \dim(\text{Im } F) \text{ and } \text{nullity}(F) = \dim(\text{Ker } F).$$

Theorem 8.6: Let V be of finite dimension, and let $F: V \to U$ be linear. Then $\dim V = \dim(\ker F) + \dim(\text{Im } F) = \text{nullity}(F) + \text{rank}(F)$.

Recall that the rank of a matrix A was also defined to be the dimension of its column space and row space. If we now view A as a linear mapping, then both definitions correspond, since the image of A is precisely its column space.

Example 8.9. Let F: $\mathbf{R}^4 \to \mathbf{R}^3$ be the linear mapping defined by

$$F(x, y, z, t) = (x - y + z + t,\ 2x - 2y + 3z + 4t,\ 3x - 3y + 4z + 5t)$$

(*a*) Find a basis and the dimension of the image of F. First find the image of the usual basis vectors of $\mathbf{R}^4$,

$$F(1, 0, 0, 0) = (1, 2, 3), \qquad F(0, 0, 1, 0) = (1, 3, 4)$$
$$F(0, 1, 0, 0) = (-1, -2, -3), \quad F(0, 0, 0, 1) = (1, 4, 5)$$

By Proposition 5.4, the image vectors span Im F. Hence form the matrix M whose rows are these image vectors and row reduce to echelon form:

$$M = \begin{bmatrix} 1 & 2 & 3 \\ -1 & -2 & -3 \\ 1 & 3 & 4 \\ 1 & 4 & 5 \end{bmatrix} \sim \begin{bmatrix} 1 & 2 & 3 \\ 0 & 0 & 0 \\ 0 & 1 & 1 \\ 0 & 2 & 2 \end{bmatrix} \sim \begin{bmatrix} 1 & 2 & 3 \\ 0 & 1 & 1 \\ 0 & 0 & 0 \\ 0 & 0 & 0 \end{bmatrix}$$

Thus (1, 2, 3) and (0, 1, 1) form a basis of Im F. Hence

$$\dim(\text{Im } F) = 2 \text{ and } \text{rank}(F) = 2.$$

(*b*) Find a basis and the dimension of the kernel of the map F. Set $F(v) = 0$, where $v = (x, y, z, t)$,

$$F(x, y, z, t) = (x - y + z + t,\ 2x - 2y + 3z + 4t,\ 3x - 3y + 4z + 5t)$$

Set corresponding components equal to each other to form the following homogeneous system whose solution space is Ker F:

$$\begin{aligned} x - y + z + t &= 0 \\ 2x - 2y + 3z + 4t &= 0 \\ 3x - 3y + 4z + 5t &= 0 \end{aligned} \quad \text{or} \quad \begin{aligned} x - y + z + t &= 0 \\ z + 2t &= 0 \\ z + 2t &= 0 \end{aligned} \quad \text{or} \quad \begin{aligned} x - y + z + t &= 0 \\ z + 2t &= 0 \end{aligned}$$

The free variables are y and t. Hence

$$\dim(\text{Ker } F) = 2 \text{ or } \text{nullity}(F) = 2.$$

(i) Set $y = 1, t = 0$ to obtain the solution $(-1, 1, 0, 0)$,
(ii) Set $y = 0, t = 1$ to obtain the solution $(1, 0, -2, 1)$.
Thus $(-1, 1, 0, 0)$ and $(1, 0, -2, 1)$ form a basis for Ker F.
As expected, $\dim(\text{Im } F) + \dim(\text{Ker } F) = 4 = \dim \mathbf{R}^4$.

Let $F: V \to U$ be a linear mapping. Recall that $F(0) = 0$. F is said to be singular if the image of some nonzero vector v is 0, that is, if there exists $v \neq 0$ such that $F(v) = 0$. Thus $F: V \to U$ is *nonsingular* if the zero vector 0 is the only vector whose image under F is 0 or, in other words, if Ker $F = \{0\}$.

Theorem 8.7: Let $F: V \to U$ be a nonsingular linear mapping. Then the image of any linearly independent set is linearly independent.

Suppose a linear mapping $F: V \to U$ is one-to-one. Then only $0 \in U$, and so F is nonsingular. The converse is also true. For suppose F is nonsingular and $F(v) = F(w)$, then $F(v - w) = F(v) - F(w) = 0$, and hence $v - w = 0$ or $v = w$. Thus $F(v) = F(w)$ implies $v = w$, that is, F is one-to-one. Thus we have proved the following proposition.

Proposition 8.8: A linear mapping $F: V \to U$ is one-to-one if and only if F is nonsingular.

Recall that a mapping $F: V \to U$ is called an *isomorphism* if F is linear and if F is bijective, i.e., if F is one-to-one and onto. Also, recall that a vector space V is said to be *isomorphic* to a vector space U, written $V \cong U$, if there is an isomorphism $F: V \to U$.

Theorem 8.9: Suppose V has finite dimension and $\dim V = \dim U$. Suppose $F: V \to U$ is linear. Then F is an isomorphism if and only if F is nonsingular.

Operations with Linear Mappings

We are able to combine linear mappings in various ways to obtain new linear mappings. These operations are very important and will be used throughout the text.

Let $F: V \to U$ and $G: V \to U$ be linear mappings over a field K. The sum $F + G$ and the scalar product kF, where $k \in K$, are defined to be the following mappings from V into U:

$$(F+G)(v) \equiv F(v) + G(v) \text{ and } (kF)(v) \equiv kF(v)$$

We now show that if F and G are linear, then $F + G$ and kF are also linear. Specifically, for any vectors $v, w, \in V$ and any scalars $a, b \in K$,

$$\begin{aligned} (F+G)(v) &= F(av+bw)+G(av+bw) \\ &= aF(v)+bF(w)+aG(v)+bG(w) \\ &= a[F(v)+G(v)]+b[F(w)+G(w)] \\ &= a(F+G)(v)+b(F+G)(w) \end{aligned}$$

and

$$\begin{aligned} (kF)(av+bw) &= kF(av+bw) = k[aF(v)+bF(w)] \\ &= akF(v)+bkF(w) = a(kF)(v)+b(kf)(w) \end{aligned}$$

Thus $F + G$ and kF are linear.

Theorem 8.10: Let V and U be vector spaces over a field K. Then the collection of all linear mappings from V into U with the above operations of addition and scalar multiplication forms a vector space over K.

The vector space of linear mappings in the above Theorem 8.10 is usually denoted by $\text{Hom}(V, U)$.

Theorem 8.11: Suppose $\dim V = m$ and $\dim U = n$. Then

$$\dim[\text{Hom}(V, U)] = mn.$$

Let V be a vector space over a field K. This section considers the special case of linear mappings from the vector space V into itself, that is, linear mappings of the form $F: V \rightarrow V$. They are also called *linear operators* or *linear transformations* on V. We will write $A(V)$, instead of Hom(V, V), for the space of all such mappings.

Now $A(V)$ is a vector space over K, and, if $\dim V = n$, then $\dim A(V) = n^2$. Moreover, for any mappings $F, G \in A(V)$, the composition $G \circ F$ exists and also belongs to $A(V)$. Thus we have a "multiplication" defined in $A(V)$. [We sometimes write GF instead of $G \circ F$ in the space $A(V)$.]

An *algebra* A over a field K is a vector space over K in which an operation of multiplication is defined satisfying, for every $F, G, H \in A$ and every $k \in K$.

(i) $F(G + H) = FG + FH$,

(ii) $(G + H)F = GF + HF$,

(iii) $k(GF) = (kG)F = G(kF)$.

The algebra is said to be *associative* if, in addition, $(FG)H = F(GH)$.

Theorem 8.12: Let V be a vector space over K. Then $A(V)$ is an associative algebra over K with respect to composition of mappings. If $\dim V = n$, then $\dim A(V) = n^2$.

This is why $A(V)$ is called the *algebra of linear operators* on V.

Linear Mappings and Matrices

Let T be a linear operator from a vector space V into itself, and suppose $S = \{u_1, u_2, \ldots, u_n\}$ is a basis of V. Now $T(u_1), T(u_2), \ldots, T(u_n)$ are vectors in V, and so each is a linear combination of the vectors in the basis S; say,

$$T(u_1) = a_{11}u_1 + a_{12}u_2 + \ldots + a_{1n}u_n$$

$$T(u_2) = a_{21}u_1 + a_{22}u_2 + \ldots + a_{2n}u_n$$

$$\cdots\cdots\cdots\cdots\cdots\cdots\cdots\cdots\cdots\cdots\cdots$$

$$T(u_n) = a_{n1}u_1 + a_{n2}u_2 + \ldots + a_{nn}u_n$$

The transpose of the above matrix of coefficients, denoted by $m_S(T)$ or $[T]_S$, is called the *matrix representation* of T relative to the basis S, or simply the matrix of T in the basis S. (The subscript S may be omitted if the basis S is understood.)

Using the coordinate (column) vector notation, the matrix representation of T may be written in the form

$$m_S(T) = [T]_S = \left[[T(u_1)]_S, [T(u_2)]_S, \ldots, [T(u_n)]_S\right]$$

That is, the columns of $m(T)$ are the coordinate vectors of $T(u_1)$, $T(u_2)$, … , $T(u_n)$, respectively.

Example 8.10. Let F: $\mathbf{R}^2 \to \mathbf{R}^2$ be the linear operator defined by $F(x, y) = (2x + 3y, 4x - 5y)$. Find the matrix representation of F relative to the basis $S = \{u_1, u_2\} = \{(1, 2), (2, 5)\}$.

(1) First find $F(u_1)$, and then write it as a linear combination of the basis vectors u_1 and u_2. (For notational convenience, we use column vectors.) We have

$$F(u_1) = F\left(\begin{bmatrix}1\\2\end{bmatrix}\right) = \begin{bmatrix}8\\-6\end{bmatrix} = x\begin{bmatrix}1\\2\end{bmatrix} + y\begin{bmatrix}2\\5\end{bmatrix} \quad \text{and} \quad \begin{matrix}x + 2y = 8\\2x + 5y = -6\end{matrix}$$

Solve the system to obtain $x = 52$ and $y = -22$.
Hence $F(u_1) = 52u_1 - 22u_2$.

(2) Next find $F(u_2)$, and then write it as a linear combination of u_1 and u_2:

$$F(u_2) = F\left(\begin{bmatrix}2\\5\end{bmatrix}\right) = \begin{bmatrix}19\\-17\end{bmatrix} = x\begin{bmatrix}1\\2\end{bmatrix} + y\begin{bmatrix}2\\5\end{bmatrix} \quad \text{and} \quad \begin{matrix}x + 2y = 19\\2x + 5y = -17\end{matrix}$$

Solve the system to obtain $x = 129$, and $y = -55$.
Thus $F(u_2) = 129u_1 - 55u_2$.

Now write the coordinates of $F(u_1)$ and $F(u_2)$ as columns to obtain the matrix $[F]_S = \begin{bmatrix}52 & 129\\-22 & -55\end{bmatrix}$

Next follows an algorithm for finding matrix representations. The first Step 0 is optional. It may be useful to use it in Step 1(b), which is repeated for each basis vector.

Algorithm 8.1: The input is a linear operator T on a vector space V and a basis $S = \{u_1, u_2, \ldots, u_n\}$ of V. The output is the matrix representation $[T]_S$.

Step 0. Find a formula for the coordinates of an arbitrary vector v relative to the basis S.

Step 1. Repeat for each basis vector u_k in S:

(*a*) Find $T(u_k)$.

(*b*) Write $T(u_k)$ as a linear combination of the basis vectors

$$u_1, u_2, \ldots, u_n.$$

Step 2. Form the matrix $[T]_S$ whose columns are the coordinate vectors in Step 1(*b*).

Example 8.11. Let $F: \mathbf{R}^2 \to \mathbf{R}^2$ be defined by $F(x, y) = (2x + 3y, 4x - 5y)$. Find the matrix representation $[F]_S$ of F relative to the basis $S = \{u_1, u_2\} = \{(1, -2), (2, -5)\}$.

(Step 0) First find the coordinates of $(a, b) \in \mathbf{R}^2$ relative to the basis S. We have

$$\begin{bmatrix} a \\ b \end{bmatrix} = x\begin{bmatrix} 1 \\ -2 \end{bmatrix} + y\begin{bmatrix} 2 \\ -5 \end{bmatrix} \quad \text{or} \quad \begin{matrix} x + 2y = a \\ -2x - 5y = b \end{matrix} \quad \text{or} \quad \begin{matrix} x + 2y = a \\ -y = 2a + b \end{matrix}$$

Solving for x and y in terms of a and b yields $x = 5a + 2b$, $y = -2a - b$. Thus
$(a, b) = (5a + 2b)u_1 + (-2a - b)u_2$

(Step 1) Now we find $F(u_1)$ and write it as a linear combination of u_1 and u_2 using the above formula for (a, b), and then we repeat the process for $F(u_2)$. We have

$$F(u_1) = F(1, -2) = (-4, 14) = 8u_1 - 6u_2$$
$$F(u_2) = F(2, -5) = (-11, 33) = 11u_1 - 11u_2$$

(Step 2) Finally, we write the coordinates of $F(u_1)$ and $F(u_2)$ as columns to obtain the required matrix

$$[F]_S = \begin{bmatrix} 8 & 11 \\ -6 & -11 \end{bmatrix}$$

Theorem 8.13: Let $T: V \to V$ be a linear operator, and let S be a (finite) basis of V. Then, for any vector v in V, $[T]_S[v]_S = [T(v)]_S$.

Given a basis S of a vector space V, we have associated a matrix $[T]$ to each linear operator T in the algebra $A(V)$ of linear operators on V. Theorem 8.13 tells us that the "action" of an individual linear operator T is preserved by this representation. The next two theorems tell us that the three basic operations in $A(V)$ with these operators, namely (i) addition, (ii) scalar multiplication, and (iii) composition, are also preserved.

Theorem 8.14: Let V be an n-dimensional vector space over K, let S be a basis of V, and let $\mathbf{M}$ be the algebra of $n \times n$ matrices over K. Then the mapping: m: $A(V) \to \mathbf{M}$ defined by $m(T) = [T]_S$ is a vector space isomorphism. That is, for any $F, G \in A(V)$ and any $k \in K$,

(i) $m(F + G) = m(F) + m(G)$ or $[F + G] = [F] + [G]$
(ii) $m(kF) = km(F)$ or $[kF] = k[F]$
(iii) m is bijective (one-to-one and onto).

Theorem 8.15: For any linear operators $F, G \in A(V)$,

$$m(G \circ F) = m(G)m(F) \text{ or } [G \circ F] = [G][F]$$

Change of Basis

Let V be an n-dimensional vector space over a field K. We have shown that, once we have selected a basis S of V, every vector $v \in V$ can be represented by means of an n-tuple $[v]_S$ in K^n, and every linear operator T in $A(V)$ can be represented by an $n \times n$ matrix over K. We ask the following natural question:
How do our representations change if we select another basis?
In order to answer this question, we first need a definition.

Let $S = \{u_1, u_2, \ldots, u_n\}$ be a basis of a vector space V, and let $S' = \{v_1, v_2, \ldots, v_n\}$ be another basis. (For reference, we will call S the "old" basis and S (the "new" basis.) Since S is a basis, each vector in the "new" basis S (can be written uniquely as a linear combination of the vectors in S; say,

$$v_1 = a_{11}u_1 + a_{12}u_2 + \ldots + a_{1n}u_n$$
$$v_2 = a_{21}u_1 + a_{22}u_2 + \ldots + a_{2n}u_n$$
$$\ldots\ldots\ldots\ldots\ldots\ldots\ldots\ldots\ldots\ldots\ldots\ldots\ldots\ldots$$
$$v_n = a_{n1}u_1 + a_{n2}u_2 + \ldots + a_{nn}u_n$$

Let P be the transpose of the above matrix of coefficients; that is, let $P = [p_{ij}]$, where $p_{ij} = a_{ij}$. Then P is called the *change-of-basis matrix* (or *transition matrix*) from the "old" basis S to the new basis S'.

The above change-of-basis matrix P may also be viewed as the matrix whose columns are, respectively, the coordinate column vectors of the "new" basis vectors v_i relative to the "old" basis S; namely,

$$P = \left[[v_1]_S, [v_2]_S, \ldots, [v_n]_S\right]$$

Analogously, there is a change-of-basis matrix Q from the "new" basis to the "old" basis S. Similarly, Q may be viewed as the matrix whose columns are, respectively, the coordinate column vectors of the "old" basis S' vectors u_i relative to the "new" basis S; namely,

$$Q = \left[[u_1]_{S'}, [u_2]_{S'}, \ldots, [u_n]_{S'}\right]$$

Since the vectors $v_1, v_2, \ldots, v_n$ in the new basis S' are linearly independent, the matrix P is invertible. Similarly, Q is invertible. In fact, we have the following proposition.

Proposition 8.16: Let P and Q be the above change-of-basis matrices. Then $Q = P^{-1}$.

Now suppose $S = \{u_1, u_2, \ldots, u_n\}$ is a basis of a vector space V, and suppose $P = [p_{ij}]$ is any nonsingular matrix. Then the n vectors $v_i = p_{1i}u_1 + p_{2i}u_2 + \ldots + p_{ni}u_n$, $i = 1, 2, \ldots, n$ corresponding to the columns of P will be the change-of-basis matrix from S to the new basis S'.

Example 8.12. Consider the following two bases of $\mathbf{R}^2$:

$$S = \{u_1, u_2\} = \{(1, 2), (3, 5)\} \text{ and } S' = \{v_1, v_2\} = \{(1, -1), (1, -2)\}$$

(*a*) Find the change-of-basis matrix P from S to the "new" basis S'
Write each of the new basis vectors of S' as a linear combination of the original basis vectors u_1 and u_2 of S. We have

$$\begin{bmatrix} 1 \\ -1 \end{bmatrix} = x\begin{bmatrix} 1 \\ 2 \end{bmatrix} + y\begin{bmatrix} 3 \\ 5 \end{bmatrix} \text{ or } \begin{matrix} x+3y=1 \\ 2x+5y=-1 \end{matrix} \text{ yielding } x=-8,\ y=3$$

$$\begin{bmatrix} 1 \\ -2 \end{bmatrix} = x\begin{bmatrix} 1 \\ 2 \end{bmatrix} + y\begin{bmatrix} 3 \\ 5 \end{bmatrix} \text{ or } \begin{matrix} x+3y=1 \\ 2x+5y=-2 \end{matrix} \text{ yielding } x=-11,\ y=4$$

Thus

$$\begin{matrix} v_1 = -8u_1 + 3u_2 \\ v_2 = -11u_1 + 4u_2 \end{matrix} \quad \text{and hence } P = \begin{bmatrix} -8 & -11 \\ 3 & 4 \end{bmatrix}$$

Note that the coordinators of v_1 and v_2 are the columns, not rows, of the change-of-basis matrix P.

(*b*) Find the change-of-basis matrix Q from the "new" basis S' back to the "old" basis S.
Here we write each of the "old" basis vectors u_1 and u_2 of S' as a linear combination of the "new" basis vectors v_1 and v_2 of S'. This yields

$$\begin{matrix} u_1 = 4v_1 - 3v_2 \\ u_2 = 11v_1 - 8v_2 \end{matrix} \quad \text{and hence } Q = \begin{bmatrix} 4 & 11 \\ -3 & -8 \end{bmatrix}$$

As expected from Proposition 8.16, $Q = P^{-1}$. (In fact, we could have obtained Q by simply finding P^{-1}.)

Example 8.13. Consider the following two bases of $\mathbf{R}^3$:

$$E = \{e_1, e_2, e_3\} = \{(1, 0, 0), (0, 1, 0), (0, 0, 1)\}$$
$$\text{and } S = \{u_1, u_2, u_3\} = \{(1, 0, 1), (2, 1, 2), (1, 2, 2)\}$$

(*a*) Find the change-of-basis matrix P from the basis E to the basis S.
Since E is the usual basis, we can immediately write each basis element of S as a linear combination of the basis elements of E. Specifically,

$$u_1 = (1,0,1) = e_1 + e_3$$
$$u_2 = (2,1,2) = 2e_1 + e_2 + 2e_3 \quad \text{and hence} \quad P = \begin{bmatrix} 1 & 2 & 1 \\ 0 & 1 & 2 \\ 1 & 2 & 2 \end{bmatrix}$$
$$u_3 = (1,2,2) = e_1 + 2e_2 + 2e_3$$

Again, the coordinates of u_1, u_2, u_3 appear as the columns in P. Observe that P is simply the matrix whose columns are the basis vectors of S. This is true only because the original basis was the usual basis E.

(*b*) Find the change-of-basis matrix Q from the basis S to the basis E.

The definition of the change-of-basis matrix Q tells us to write each of the (usual) basis vectors in E as a linear combination of the basis elements of S. This yields

$$e_1 = (1,0,1) = -2u_1 + 2u_2 - u_3$$
$$e_2 = (0,1,0) = -2u_1 + u_2 \quad \text{and hence} \quad Q = \begin{bmatrix} -2 & -2 & 3 \\ 2 & 1 & -2 \\ -1 & 0 & 1 \end{bmatrix}$$
$$e_3 = (0,0,1) = 3u_1 - 2u_2 + u_3$$

We emphasize that to find Q, we need to solve three 3×3 systems of linear equations – one 3×3 system for each of e_1, e_2, e_3.

Alternately, we can find $Q = P^{-1}$ by forming the matrix $M = [P, I]$ and row reducing M to row canonical form:

$$M = \begin{bmatrix} 1 & 2 & 1 & 1 & 0 & 0 \\ 0 & 1 & 2 & 0 & 1 & 0 \\ 1 & 2 & 2 & 0 & 0 & 1 \end{bmatrix} \sim \begin{bmatrix} 1 & 0 & 0 & -2 & -2 & 3 \\ 0 & 1 & 0 & 2 & 1 & -2 \\ 0 & 0 & 1 & -1 & 0 & 1 \end{bmatrix} = [I, P^{-1}]$$

$$\text{thus } Q = P^{-1} = \begin{bmatrix} -2 & -2 & 3 \\ 2 & 1 & -2 \\ -1 & 0 & 1 \end{bmatrix}$$

(Here we have used the fact that Q is the inverse of P.)

Proposition 8.17: The change-of-basis matrix from the usual basis E of K^n to any basis S of K^n is the matrix P whose columns are, respectively, the basis vectors of S.

Matrices and General Linear Mappings

Lastly, we consider the general case of linear mappings from one vector space into another. Suppose V and U are vector spaces over the same field K and, say, $\dim V = m$ and $\dim U = n$. Furthermore, suppose $S = \{v_1, v_2, \ldots, v_n\}$ and $S' = \{u_1, u_2, \ldots, u_n\}$ are arbitrary but fixed bases, respectively, of V and U.

Suppose $F: V \to U$ is a linear mapping. Then the vectors $F(v_1)$, $F(v_2)$, $\ldots$, $F(v_m)$ belong to U, and so each is a linear combination of the basis vectors in S'; say,

$$F(v_1) = a_{11}u_1 + a_{12}u_2 + \ldots + a_{1n}u_n$$
$$F(v_2) = a_{21}u_1 + a_{22}u_2 + \ldots + a_{2n}u_n$$
$$\ldots\ldots\ldots\ldots\ldots\ldots\ldots\ldots\ldots\ldots\ldots\ldots$$
$$F(v_m) = a_{m1}u_1 + a_{m2}u_2 + \ldots + a_{mn}u_n$$

The transpose of the above matrix of coefficients, denoted by $m_{S,S'}(F)$ or $[F]_{S,S'}$, is called the *matrix representation* of F relative to the bases S and S'. [We will use the simple notation $m(F)$ and $[F]$ when the bases are understood.]
The following theorem is analogous to Theorem 8.13 for linear operators.

Theorem 8.18: For any vector $v \in V$, $[F]_{S,S'}[v]_S = [F(v)]_{S'}$.

That is, multiplying the coordinates of v in the basis S of V by $[F]$, we obtain the coordinates of $F(v)$ in the basis S' of U.

Recall that for any vector spaces V and U, the collection of all linear mappings from V into U is a vector space and is denoted by $\mathrm{Hom}(V, U)$. The following theorem is analogous to Theorem 8.14 for linear operators, where now we let $\mathbf{M} = \mathbf{M}_{m,n}$ denote the vector space of all $m \times n$ matrices.

Theorem 8.19: The the mapping: m: $\mathrm{Hom}(V, U) \to \mathbf{M}$ defined by $m(F) = [F]_S$ is a vector space isomorphism. That is, for any $F, G \in \mathrm{Hom}(V, U)$ and any scalar k,

(i) $m(F + G) = m(F) + m(G)$ or $[F + G] = [F] + [G]$
(ii) $m(kF) = km(F)$ or $[kF] = k[F]$
(iii) m is bijective (one-to-one and onto).

Our next theorem is analogous to Theorem 8.15 for linear operators.

Theorem 8.20: Let S, S', S'' be bases of vector spaces V, U, W, respectively. Let $F{:}V \to U$ and G: $U \to W$ be linear mappings. Then

$$[G \circ F]_{S', S''} = [G]_{S', S''} [F]_{S', S''}$$

That is, relative to the appropriate bases, the matrix representation of the composition of two mappings is the matrix product of the matrix representation of the individual mappings.

Next we show how the matrix representation of the linear mapping F: $V \to U$ is affected when new bases are selected.

Theorem 8.21: Let P be the change-of-basis matrix from a basis e to a basis e' in V, and let Q be the change-of-basis matrix from a basis f to a basis f' in U. Then, for any linear map F: $V \to U$,

$$[F]_{e',f'} = Q^{-1}[F]_{e,f}P$$

In other words, if A is the matrix representation of a linear mapping F relative to the bases e and f, and B is the matrix representation of F relative to the bases e' and f', then $B = Q^{-1}AP$.

Our last theorem shows that any linear mapping from one vector space V into another vector space U can be represented by a very simple matrix.

Theorem 8.22: Let F: $V \to U$ be linear and, say, rank$(F) = r$. Then there exist bases of V and U such that the matrix representation of F has the form $A = \begin{bmatrix} I_r & 0 \\ 0 & 0 \end{bmatrix}$ where I_r is the r-square identity matrix.

The above matrix A is called the *normal* or *canonical form* of the linear map F.

Chapter 9
CANONICAL FORMS

IN THIS CHAPTER:

- ✔ *Triangular Form*
- ✔ *Invariance*
- ✔ *Invariant Direct-Sum Decompositions*
- ✔ *Primary Decomposition*
- ✔ *Nilpotent Operators*
- ✔ *Jordan Canonical Form*
- ✔ *Cyclic Subspaces*
- ✔ *Rational Canonical Form*

Triangular Form

Let T be a linear operator on an n-dimensional vector space V. Suppose T can be represented by the triangular matrix

$$A = \begin{bmatrix} a_{11} & a_{12} & \dots & a_{1n} \\ & a_{22} & \dots & a_{2n} \\ & & \dots & \dots \\ & & & a_{nn} \end{bmatrix}$$

Then the characteristic polynomial $\Delta(t)$ of T is a product of linear factors; that is,

$$\Delta(t) = \det(tI - A) = (t - a_{11})\,(t - a_{22})\,\ldots\,(t - a_{nn})$$

The converse is also true and is an important theorem.

Theorem 9.1: Let $T{:}V \to V$ be a linear operator whose characteristic polynomial factors into linear polynomials. Then there exists a basis of V in which T is represented by a triangular matrix.

Theorem 9.1: (**Alternative Form**) Let A be a square matrix whose characteristic polynomial factors into linear polynomials. Then A is similar to a triangular matrix, i.e., there exists an invertible matrix P such that $P^{-1}AP$ is triangular.

We say that an operator T can be brought into triangular form if it can be represented by a triangular matrix. Note that in this case, the eigenvalues of T are precisely those entries appearing on the main diagonal. We give an application of this remark.

Example 9.1. Let A be a square matrix, over the complex field **C**. Suppose λ is an eigenvalue of A^2. Show that $\sqrt{\lambda}$ or $-\sqrt{\lambda}$ is an eigenvalue of A.

By Theorem 9.1, A and A^2 are similar, respectively, to triangular matrices of the form

$$B = \begin{bmatrix} \mu_1 & * & \cdots & * \\ & \mu_2 & \cdots & * \\ & & \cdots & \cdots \\ & & & \mu_n \end{bmatrix} \quad \text{and} \quad B^2 = \begin{bmatrix} \mu_1^2 & * & \cdots & * \\ & \mu_2^2 & \cdots & * \\ & & \cdots & \cdots \\ & & & \mu_n^2 \end{bmatrix}$$

Since similar matrices have the same eigenvalues, $\lambda = \mu_i^2$ for some i. Hence $\mu_i = \sqrt{\lambda}$ or $\mu_i = -\sqrt{\lambda}$ is an eigenvalue of A.

Invariance

Let $T{:}V \to V$ be linear. A subspace W of V is said to be *invariant under* T or *T-invariant* if T maps W into itself, i.e., if $v \in W$ implies $T(v) \in W$. In this case, T restricted to W defines a linear operator on W; that is, T induces a linear operator $\hat{T} : W \to W$ defined by $\hat{T}(w) = T(w)$ for every $w \in W$.

Example 9.2. Nonzero eigenvectors of a linear operator $T:V \to V$ may be characterized as generators of T-invariant 1-dimensional subspaces. For example, suppose $T(v) = \lambda v$, $v \neq 0$. Then $W = \{kv, k \in K\}$, the 1-dimensional subspace generated by v, is invariant under T because

$$T(kv) = kT(v) = k(\lambda v) = k\lambda v \in W$$

Conversely, suppose $\dim U = 1$ and $u \neq 0$ spans U, and U is invariant under T. Then $T(u) \in U$ and so $T(u)$ is a multiple of u, i.e., $T(u) = \mu u$. Hence u is an eigenvector of T.

The next theorem gives us an important class of invariant subspaces.

Theorem 9.2: Let $T:V \to V$ be any linear operator, and let $f(t)$ be any polynomial. Then the kernel of $f(T)$ is invariant under T.

The notion of invariance is related to matrix representations as follows.

Theorem 9.3: Suppose W is an invariant subspace. Let $T:V \to V$. Then T has a block matrix representation $\begin{bmatrix} A & B \\ 0 & C \end{bmatrix}$, where A is a matrix representation of the restriction $\hat{T}$ of T to W.

Invariant Direct-Sum Decompositions

A vector space V is termed the *direct sum* of subspaces $W_1, \ldots, W_r$, written $V = W_1 \oplus W_2 \oplus \ldots \oplus W_r$ if every vector $v \in V$ can be written uniquely in the form $v = w_1 + w_2 + \ldots + w_r$, with $w_i \in W_i$

Theorem 9.4: Suppose $W_1, W_2, \ldots, W_r$ are subspaces of V, and suppose $B_1 = \{w_{11}, w_{12}, \ldots, w_{1n_1}\}, \ldots, B_r = \{wr_1, wr_2, \ldots, w_{rn_r}\}$ are bases of W_1, $W_2, \ldots, W_r$, respectively. Then V is the direct sum of the W_i if and only if the union $B = B_1 \cup \ldots \cup B_r$ is a basis of V.

Now suppose $T:V \rightarrow V$ is linear and V is the direct sum of (nonzero) T-invariant subspaces $W_1, W_2, \ldots, W_r$; that is,

$$V = W_1 \oplus W_2 \oplus \ldots \oplus W_r \text{ and } T(W_i) \subseteq W_i,\ i = 1, \ldots, r$$

Let T_i denote the restriction of T to W_i. Then T is said to be *decomposable* into operators T_i or T is said to be the *direct sum* of the T_i, written $T = T_1 \oplus T_2 \oplus \ldots \oplus T_r$. Also, the subspaces $W_1, W_2, \ldots, W_r$ are said to *reduce T* or to form a *T-invariant direct-sum decomposition* of V.

Consider the special case where two subspaces U and W reduce an operator $T:V \rightarrow V$; say $\dim U = 2$ and $\dim W = 3$ and suppose $\{u_1, u_2\}$ and $\{w_1, w_2, w_3\}$ are bases of U and W, respectively. If T_1 and T_2 denote the restrictions of T to U and W, respectively, then

$$\begin{aligned} T_1(u_1) &= a_{11}u_1 + a_{12}u_2 \\ T_1(u_2) &= a_{21}u_1 + a_{22}u_2 \end{aligned} \qquad \begin{aligned} T(w_1) &= b_{11}w_1 + b_{12}w_2 + b_{13}w_3 \\ T(w_2) &= b_{21}w_1 + b_{22}w_2 + b_{23}w_3 \\ T(w_3) &= b_{31}w_1 + b_{32}w_2 + b_{33}w_3 \end{aligned}$$

Accordingly, the following matrices A, B, M are the matrix representations of T_1, T_2, T, respectively:

$$A = \begin{bmatrix} a_{11} & a_{21} \\ a_{12} & a_{22} \end{bmatrix}, \quad B = \begin{bmatrix} b_{11} & b_{21} & b_{31} \\ b_{12} & b_{22} & b_{32} \\ b_{13} & b_{23} & b_{33} \end{bmatrix}, \quad M = \begin{bmatrix} A & 0 \\ 0 & B \end{bmatrix}$$

The block diagonal matrix M results from the fact that $\{u_1, u_2, w_1, w_2, w_3\}$ is a basis of V, and that $T(u_i) = T_1(u_i)$ and $T(w_j) = T_2(w_j)$.

A generalization of the above argument gives us the following theorem.

Theorem 9.5: Suppose $T:V \rightarrow V$ is linear and suppose V is the direct sum of T-invariant subspaces, say, $W_1, \ldots, W_r$. If A_i is a matrix representation of the restriction of T to W_i, then T can be represented by the block diagonal matrix $M = \operatorname{diag}(A_1, A_2, \ldots, A_r)$.

Primary Decomposition

The following theorem shows that any operator $T:V \rightarrow V$ is decomposable into operators whose minimum polynomials are powers of irreducible polynomials. This is the first step in obtaining a canonical form for T.

Theorem 9.6: **(Primary Decomposition Theorem)** Let $T:V \to V$ be a linear operator with minimal polynomial

$$m(t) = f_1(t)^{n_1} f_2(t)^{n_2} \ldots f_r(t)^{n_r}$$

where the $f_i(t)$ are distinct monic irreducible polynomials. Then V is the direct sum of T-invariant subspaces $W_1, \ldots, W_r$ where W_i is the kernel of $f_i(T)^{n_i}$ is the minimal polynomial of the restriction of T to W_i.

The above polynomials $f_i(t)^{n_i}$ are relatively prime. Therefore, the above fundamental theorem follows from the next two theorems.

Theorem 9.7: Suppose $T:V \to V$ is linear, and suppose $f(t) = g(t)h(t)$ are polynomials such that $f(T) = \mathbf{0}$ and $g(t)$ and $h(t)$ are relatively prime. Then V is the direct sum of the T-invariant subspace U and W, where $U = \text{Ker } g(t)$ and $W = \text{Ker } h(T)$. Furthermore, if $f(\text{t})$ is the minimal polynomial of T [and $g(t)$ and $h(t)$ are monic], then $g(t)$ and $h(t)$ are the minimal polynomials of the restrictions of T to U and W, respectively.

Theorem 9.8: A linear operator $T:V \to V$ is diagonalizable if and only if its minimal polynomial $m(t)$ is a product of distinct linear polynomials.

Theorem 9.8 (Alternative Form) A matrix A is similar to a diagonal matrix if and only if its minimal polynomial is a product of distinct linear polynomials.

Example 9.3. Suppose $A \neq I$ is a square matrix for which $A^3 = I$. Determine whether or not A is similar to a diagonal matrix if A is a matrix over: (*i*) the real field $\mathbf{R}$, (*ii*) the complex field $\mathbf{C}$.

Since $A^3 = I$, A is a zero of the polynomial $f(t) = t^3 - 1 = (t - 1)(t^2 + t + 1)$. The minimal polynomial $m(t)$ of A cannot be $t - 1$, since $A \neq I$. Hence $m(t) = t^2 + t + 1$ or $m(t) = t^3 - 1$. Since neither polynomial is a product of linear polynomials over $\mathbf{R}$, A is not diagonalizable over $\mathbf{R}$. On the other hand, each of the polynomials is a product of distinct linear polynomials over $\mathbf{C}$. Hence A is diagonalizable over $\mathbf{C}$.

Nilpotent Operators

A linear operator $T:V \rightarrow V$ is termed *nilpotent* if $T^n = \mathbf{0}$ for some positive integer n; we call k the *index of nilpotency* of T if $T^k = \mathbf{0}$ but $T^{k-1} \neq \mathbf{0}$. Analogously, a square matrix A is termed nilpotent if $A^n = 0$ for some positive integer n, and of index k if $A^k = 0$ but $A^{k-1} \neq 0$. Clearly the minimum polynomial of a nilpotent operator (matrix) of index k is $m(t) = t^k$; hence 0 is its only eigenvalue.

Example 9.4. The following two r-square matrices will be used through the chapter:

$$N = N(r) = \begin{bmatrix} 0 & 1 & 0 & \dots & 0 & 0 \\ 0 & 0 & 1 & \dots & 0 & 0 \\ \dots & \dots & \dots & \dots & \dots & \dots \\ 0 & 0 & 0 & \dots & 0 & 1 \\ 0 & 0 & 0 & \dots & 0 & 0 \end{bmatrix}$$

and

$$J(\lambda) = \begin{bmatrix} \lambda & 1 & 0 & \dots & 0 & 0 \\ 0 & \lambda & 1 & \dots & 0 & 0 \\ \dots & \dots & \dots & \dots & \dots & \dots \\ 0 & 0 & 0 & \dots & \lambda & 1 \\ 0 & 0 & 0 & \dots & 0 & \lambda \end{bmatrix}$$

The first matrix N, called a *Jordan nilpotent block*, consists of 1's above the diagonal (called the *superdiagonal*), and 0's elsewhere. It is a nilpotent matrix of index r. (The matrix N of order 1 is just the 1×1 zero matrix [0].)

The second matrix $J(\lambda)$, called a *Jordan block* belonging to the eigenvalue λ, consists of λ's on the diagonal, 1's on the superdiagonal, and 0's elsewhere. Observe that $J(\lambda) = \lambda I + N$. In fact, we will prove that any linear operator T can be decomposed into operators, each of which is the sum of a scalar operator and a nilpotent operator.

Theorem 9.9: Let T: $V \rightarrow V$ be a nilpotent operator of index k. Then T has a block diagonal matrix representation in which each diagonal entry is a Jordan nilpotent block N. There is at least one N of order k, and all other N are of orders $\leq k$. The number of N of each possible order is uniquely determined by T. The total number of N of all orders is equal to the nullity of T.

Jordan Canonical Form

An operator T can be put into Jordan canonical form if its characteristic and minimal polynomials factor into linear polynomials. This is always true if K is the complex field **C**. In any case, we can always extend the base field K to a field in which the characteristic and minimal polynomials do factor into linear factors; thus, in the broad sense, every operator has a Jordan canonical form. Analogously, every matrix is similar to a matrix in Jordan canonical form.

The following theorem describes the *Jordan canonical form J* of a linear operator T.

Theorem 9.10: Let T: $V \to V$ be a linear operator whose characteristic and minimal polynomials are, respectively,

$$\Delta(t) = (t - \lambda_1)^{n_1} \ldots (t - \lambda_r)^{n_r} \text{ and } m(t) = (t - \lambda_1)^{m_1} \ldots (t - \lambda_r)^{m_r}$$

where the λ_i are distinct scalars. Then T has a block diagonal matrix representation J in which each diagonal entry is a Jordan block $J_{ij} = J(\lambda_i)$. For each λ_{ij}, the corresponding J_{ij} have the following properties:

(i) There is at least one J_{ij} of order m_i; all other J_{ij} are of order less than or equal to m_i.
(ii) The sum of the orders of the J_{ij} is n_i.
(iii) The number of J_{ij} equals the geometric multiplicity of λ_i.
(iv) The number of J_{ij} of each possible order is uniquely determined by T.

Example 9.5. Suppose the characteristic and minimal polynomials of an operator T are, respectively,

$$\Delta(t) = (t-2)^4(t-5)^3 \text{ and } m(t) = (t-2)^2(t-5)^3$$

Then the Jordan canonical form of T is one of the following block diagonal matrices:

$$\mathit{diag}\left(\begin{bmatrix} 2 & 1 \\ 0 & 2 \end{bmatrix}, \begin{bmatrix} 2 & 1 \\ 0 & 2 \end{bmatrix}, \begin{bmatrix} 5 & 1 & 0 \\ 0 & 5 & 1 \\ 0 & 0 & 5 \end{bmatrix}\right) \text{ or } \mathit{diag}\left(\begin{bmatrix} 2 & 1 \\ 0 & 2 \end{bmatrix}, [2], [2], \begin{bmatrix} 5 & 1 & 0 \\ 0 & 5 & 1 \\ 0 & 0 & 5 \end{bmatrix}\right)$$

The first matrix occurs if T has two independent eigenvectors belonging to the eigenvalue 2; and the second matrix occurs if T has three independent eigenvectors belonging to 2.

Appendix
LIST OF SYMBOLS

Index